Primary route network

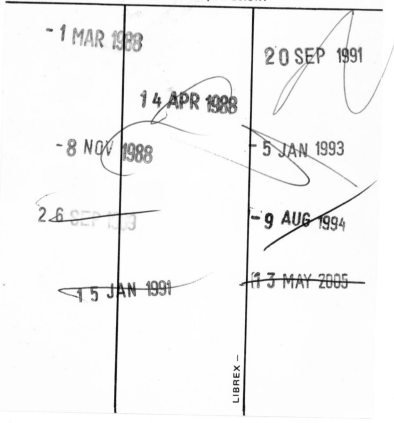

Primary route network

Proceedings of the conference 'Further developments for a primary route network', organized by The Institution of Civil Engineers and held in London on 4–5 March 1986

T Thomas Telford, London

Organizing Committee:

M. F. Hardy, Hertfordshire County Council (Chairman)
J. R. Coates, Department of Transport
A. E. Naylor, Greater Manchester Council
F. J. Parker, W. S. Atkins & Partners
R. K. Turner, Freight Transport Association

First published 1986

British Library Cataloguing in Publication Data

Further Developments for a Primary Route Network (Conference : 1986 : London)
 Primary Route Network : proceedings of a conference "Further Developments for a Primary Route Network".
 1. Road construction 2. Roads — Maintenance and repair
 I. Title   II. Institution of Civil Engineers
 625.7   TE145

ISBN 0 7277 0356 0

© The Institution of Civil Engineers, 1986, unless otherwise stated.

All rights, including translation, reserved. Except for fair copying no part of this publication may be reproduced, stored in a retrieval system, or transmitted in any form or by any means electronic, mechanical, photocopying, recording or otherwise, without the prior written permission of the publisher. Requests should be directed to the Publications Manager, Thomas Telford Ltd, P.O. Box 101, 26–34 Old Street, London EC1P 1JH.

Papers or other contributions and the statements made or opinions expressed therein are published on the understanding that the author of the contribution is solely responsible for the opinions expressed in it and that its publication does not necessarily imply that such statements and/or opinions are or reflect the views or opinions of the ICE Council or ICE committees.

Published for the Institution of Civil Engineers by Thomas Telford Ltd, P.O. Box 101, 26–34 Old Street, London EC1P 1JH.

Printed and bound in Great Britain by
Robert Hartnoll (1985) Ltd, Bodmin, Cornwall

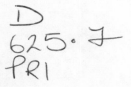

Contents

Opening Address. P. BOTTOMLEY	1
Introduction. H. W. A. FRANCIS	5

Industrial view

1. Access for retail markets. D. A. QUARMBY	9
2. Road links for chemical exports. D. J. BRICKNELL	21
Discussion on Papers 1–2	31

Central and local government policies

3. An overall review of the national network. P. E. BUTLER	35
4. A review of the primary route network within the UK. B. OLDRIDGE	49
Discussion on Papers 3–4	57

The European view

5. The development of France's primary route network over 25 years. A. THIEBAULT	69

Operational aspects

6. Electronics for dynamic traffic management: the Dutch connection. J. J. KLIJNHOUT	85
Discussion on Papers 5–6	93

Major missing links

7. The user's view. R. K. TURNER	97
8. Examples from the South East. J. S. DAWSWELL	115
9. The Ipswich bypass, western section — a case study. L. M. ELLISS	125
Discussion on Papers 7–9	137

Conurbation issues

10. Assessment of urban roads. T. E. H. WILLIAMS	145
11. Primary routes in urban areas. S. N. MUSTOW	159
Discussion on Papers 10–11	177

Opening Address

PETER BOTTOMLEY, Minister for Roads and Traffic

One of the most important aspects of road construction policy over the last few years has been the bypass programme, which has given relief to three-quarters of the historic towns and villages situated, at present, on the trunk routes. The reviewing of the primary route network is important, and changes to bring the signing up to date are shortly to be announced. In this context, I welcome the initiative taken by both the County Surveyors' Society and the Eastern Region. I think we ought to go on looking for demonstrations of partnership between central and local government.

It must be emphasised that the aim is to assess the future investment needs in general terms, and that estimates of future investment cannot be linked to a specific future network. On the National Roads front, I think we have a good story to tell without being complacent - in the last two years, we have spent within 1 per cent and 1/2 per cent respectively for the resources available. I want to pay a tribute to all those who have made that possible. We are talking about 34 road schemes under construction, with a works cost of £0.5 billion; and 282 schemes in the forward programme, with a works cost of £3.4 billion, which will add just over a thousand miles to the National Network by the late 1990s. We hope to continue to meet changing traffic demands.

By the mid-1990s, the bypass programme - that is, 220 bypasses in the period from 1979 - will have resulted in 89 per cent of towns of historic interest on trunk roads being relieved of through traffic. We are doing all we can to encourage capital investment by local highway authorities, both in their sections of the primary route network and in other major urban and rural roads. The change in the transport supplementary grant (TSG) system last year concentrated support where it could do most national good. The signs are that this is being effective in increasing the amount local authorities devote to capital expenditure on roads.

The settlement for 1986/7, announced in December 1985, gives £164 million worth of grant in support of 310 major roads schemes, with expenditure in the year of £328 million; of these schemes, two-thirds are on the primary route network.

OPENING ADDRESS

We are also talking about the 214 bypass or relief roads, of which 153 are on the primary route network. We want to work more closely with local authorities on their roads programmes for transport supplementary grant support. We are looking for long-term programmes to give a better context to bids and to help planning. the discussion of schemes in the course of preparation, and advice on suitability for TSG support. I hope we will build confidence for forward planning through the discussion of schemes in the course of preparation, advice on suitability for TSG support, and conditional acceptance so that advance costs, especially on land, can attract TSG. It might eventually be possible to work towards forward programming similar to that for trunk roads, although it is too early to say whether this is feasible. It is not enough just to assess needs, we have also to demonstrate the value of the investment in roads compared with other services. For local roads, in particular, the only measure of needs now are the bids in the TPPs.

For the next financial year, the overall increase of 10 per cent on planned spending on national and local roads includes 15 per cent for local road maintenance. This will support the speed-up of the national road construction programme in the next two to three years, and hopefully help clear the backlog of renewal work on motorways and trunk roads in about six years. I hope that local highway authorities will take the opportunity to improve construction and maintenance, especially, but not exclusively, on their part of the primary route network.

Financial resources are not the only important aspect, however. Both statutory and administrative procedures take time. It is essential to safeguard the right of the public, especially on consultation, but we need to review the administrative procedures in the Department to see whether any speed-up is possible. We would urge local authorities to take full account of procedures in timing their programmes. We are determined to press ahead with well-planned programmes essential to attaining good results on the ground. If the primary route network is to be effective, both the Department and local highway authorities as partners must make every effort to achieve good results.

It is not enough to build and maintain roads, we need also to manage the system effectively once it is in place. This means efficient gathering and dissemination of information, warnings of hazards such as fog, accidents and breakdowns, swift reactions to incidents. Ideally, there should readily be available alternative routes which are up to standard to cope with diversions. This may, however, remain an ideal for which we are striving. We need more accurate traffic forecasting in order to help keep the roads provided up to a standard capable of meeting the use demanded of them, within the resources available. We have, after all, the vital consideration of fewer accidents.

In the European Road Safety year, we need to remind

OPENING ADDRESS

ourselves constantly of the cost in both human suffering and financial terms of avoidable accidents. It is worth remembering, for example, the number of 17 and 18-year olds who are killed each year on motorbikes; or that in Northern Ireland, in the period since 1969, terrorism has claimed the lives of about 2500 people, while traffic accidents have killed about 4500 people. We give full support to European Road Safety year. It is well known that among the European Community's proposals for research is that of an investigation into road profile standards and junction layout analysed by road safety criteria.

We know that road construction improvement can yield substantial reductions in terms of accidents and we are aware of the low accident rates on motorways. One of these days, I would like to release the figures of what would happen if the traffic flows on the M25 were actually taking place on lower standard roads and thus to see whether the popular criticism of the M25 is in fact justified. COBA suggests, for example, that typical bypass schemes, which transfer traffic from an urban road to a new single carriageway road, can reduce injury accident rates by more than half. It is estimated that, in the next few years, newly completed trunk road schemes should save about 1200 injury accidents a year. A recent study showed that eleven trunk road schemes, mainly small town bypasses, which were completed between 1981-83 had an estimated reduction in accident costs of about 25 per cent. That both speaks for itself and sets the scene for this conference.

Introduction

H. W. A. FRANCIS, CBE, Vice President, The Institution of Civil Engineers

Infrastructure is now a buzz-word and infrastructure planning is almost synonymous with a plea for further investment in new construction and repair and maintenance. However, this is not the attitude of the Institution of Civil Engineers, which sees a need for strategic infrastructure planning on an economic basis and a recognition of social needs as being vitally necessary in the national interest.

The Institution started its dialogue on infrastructure and expressed the need for strategic infrastructure planning in 1974, following a period of excessive demand by the public sector on the construction industry. It was this attempt to inspire economic recovery from the public sector that led to rampant inflation. The Institution had previously expressed grave concern at the level of public sector orders for construction issued in 1972; some two-thirds more in real terms than 1984. It initiated a discussion in 1974 on: 'Civil Engineering - How much can the country afford?' at which concern was expressed at the stop/go policies of successive governments, the use of the construction industry as an economic regulator, and at the lack of strategic infrastructure planning. That concern is surely understood today - some twelve years later. I make this point to demonstrate that the Institution is not a lobby body seeking an increased allocation of public funds for construction. It is a professional body and acts accordingly.

Our members are concerned with the planning, design, construction, operation and maintenance of our national infrastructure and we are therefore in a unique position to place proposals before government on the level of investment that should be made in infrastructure in the national interest.

The Institution, in addition to initiating that early dialogue, actually proposed to the Prime Minister of the day in 1975 and 1980 that there should be a strategic infrastructure planning body to assess and plan for the national needs. It is fair to say that the Institution had a negative response on each occasion and the Council decided, therefore, in July 1981, to go it alone.

The Infrastructure Planning Group (IPG) was set up in 1981,

INTRODUCTION

with the terms of reference to review, establish and recommend infrastructure needs of the United Kingdom for the remainder of the century. The IPG's first report was published in January 1984 and is now regarded as a valuable reference document which presents an overall picture of the basic social and industrial infrastructure. The IPG then started on its second report and, in spite of the pressing problems of urban regeneration, housing, industrial development and the energy sector, it was decided that this second report should concentrate on the road sector and on water. Over these last two years, therefore, the IPG has been reviewing the state of our infrastructure investment in the transport field and particularly in the road sector. During this time, the Institution has sponsored a number of public national debates on subjects such as the Channel link, airports, energy, the water industry, railways and other issues.

This conference provides a forum for discussing further developments for a primary route network, and it is another major step forward in the Institution's programme for promoting public debate on how we might pursue the major needs of our infrastructure. The Authors have presented excellent papers, which provide an ideal backcloth for an in depth examination of the principles behind the development of a national primary route network. I know that the discussion will encompass the state of the existing assets and the operational requirements of the national infrastructure. However, we must also review the strategic needs of the long term, looking at the justification for investment and comparing our policies with those of our overseas competitors.

The IPG has based its views on three fundamental questions: 'Where are we now?'; 'Where should we be going?'; and 'How do we get there?'. The 'where are we now?' has, in reality, highlighted a weakness in the ability to define the current state of our infrastructure assets. We therefore commend the action which was initiated by Eastern Region in the reviewing of the state of infrastructure and the primary route network assets. We support the national road maintenance survey but it should be extended in its coverage, developed in its reliability, and be more widely and consistently applied by highway authorities. It is essential that we know 'where we are now'.

The Department's proposals for the collection of data and the wider adaptation of standards to the state of roads need to proceed urgently. From the data examined by the IPG, there is greater cause for concern about the state of our roads than the National Road Maintenance Survey suggests. The 1985 survey is due for publication shortly and should give some indication as to whether this view is correct. A backlog of maintenance work is building up and, on the basis of the present levels of maintenance, it will grow and not diminish. The IPG accepts the engineering picture of the overall deterioration in the state of all categories of road but regrets the lack of more reliable data. We also note with concern the results of the residual life tests, which highlight a significant percentage of classes of road with less than ten years of life; and also the shift of local authority resources towards top dressing of roads at the expense of more longer-term remedial work, arising no doubt from fincancial constraints which enforce a make do and mend policy leading inevitably to higher costs later. We also view with concern the inadequacy of infrastructure investment relative to our European competitors. Our level of investment in infrastructure relative to the major EEC countries is six out of six. The average investment in those six countries per head of population is £125. In West Germany, £209 per head of population is spent on infrastructure investment. In the United Kingdom, by contrast, expenditure is £65 per head of population.

It is important to take a strategic view and make a long-term assessment of future demand and need. The dramatic increase in exports as a percentage of gross domestic product, the shift of exports to Europe and the construction of the fixed channel link, all impinge on the assessment of the needs of the primary route network. Are we looking far enough ahead to the next changes in demand? This Conference presents us with the opportunity to debate, and to initiate action on, the three fundamental issues:

1. Where are we now? An audit of the state of our existing assets relative to need is urgently required.
2. Where should we be going? What affects the future demand? How can we assess those demand patterns?
3. How do we get there? What should be the priorities and the levels of investment in the construction of new roads, and in reconstruction, repair and maintenance?

We also have the opportunity at this conference to provide professional leadership, and to express neither subjective nor political, but highly professional views on where our primary route network should be going.

1. Access for retail markets

D. A. QUARMBY, MA, PhD, FCIT, MIRTE, Distribution Director,
J. Sainsbury plc

SYNOPSIS

This paper examines the impact that improvements in the primary road network have had and can have on the physical distribution of goods to retail markets. The first part describes recent and current trends in physical distribution to retail outlets - resources, structure and industry relationships. The second part focusses on the road transport content of physical distribution, and the relevance of the primary road network. The third part sets out to identify the type of impacts that major road network improvements can have on costs, depot structure and distribution strategy, with particular reference to Sainsbury's recent experience. It is argued that conventional appraisal methods of valuing the benefits to commercial vehicles of major highway improvements understate the "business potential" which they release.

TRENDS IN PHYSICAL DISTRIBUTION

1. Distribution of consumer goods to retail outlets - particularly food - has undergone immense change in the last twenty years. Apart from increases in the maximum gvw of commercial vehicles, the sources of change have not primarily been technological. They derive from changes in retailing itself - particularly the growth of large multiple chains both in food and in other consumer products - and from changes in the organisation and structure of the physical distribution industry.

2. The two principal trends in organisation and structure have been and continue to be

    i) away from supplier-controlled distribution networks towards retailer-controlled networks.

    ii) away from own-account operations to contract distribution.

INDUSTRIAL VIEW

To illustrate the first trend: a supplier-controlled network will typically involve trunk delivery from the supplier's manufacturing or processing factory to a large number of his distribution centres, each serving a specific area. From there, the supplier provides local delivery direct to the customers' retail premises. Sometimes there are two tiers of distribution centre or depot, but one tier will typically only be for transhipment of assembled orders. By contrast, a retailer-controlled network is one in which the retailer controls the distribution centres - which will tend to be regional in scale, larger in capacity and fewer in number than in the supplier's network. Suppliers still deliver into the distribution centre; the retailer is then responsible for the goods and for their distribution to his own retail outlets.

3. Why is this trend towards retailer-control happening? There is a number of reasons concerning the management of a retail business - flexibility to respond to rapid changes in volume and distribution, in order/delivery cycles, and in seasonal patterns and to external factors such as supplier problems, industrial relations factors, and so on. But there is also a very simple reason: it is more cost efficient. At a retail-controlled depot the economics of load consolidation, high vehicle fill and only one or two drops per vehicle is bound to incur lower resource costs than the supplier carrying out multi-drops of much smaller individual branch orders.

4. In 1982, the volume of distribution into all types of retail outlets was almost equally divided between the three main participants (Ref 1):

	%
retail-controlled distribution	35
delivered wholesaler	33
cash and carry	3
supplier-controlled distribution	30

Proportions are expected move further in favour of retail-controlled distribution, with 44% forecast for 1990, at the expense of supplier-controlled; at the same time, the delivered wholesaler, meeting the needs of small and medium sized outlets, is likely to decline in favour of direct deliveries to such outlets by suppliers. For a few years now, Tesco, and more recently Argyll, have been moving from direct supplier delivery to their own depot distribution - part contract, part own-account. Sainsbury's has depended on a high level of depot distribution for decades, now running at over 80% of volume sales.

PAPER 1: QUARMBY

5. An indication of the relative efficiency of retail-controlled systems is given by the following indices of unit costs of transport, stockholding and warehousing

    retail-controlled distribution   100
    supplier-controlled distribution   190
    delivered wholesaler   250

6. The overall cost of the physical distribution chain from suppliers' factories to all types of retail outlets was £12bn in 1982 (excluding supply to the catering and licensed trades). This represents nearly 18% of retail sales value. Typically, total distribution costs in fast moving consumer goods, and particularly food, are a lower percentage than this, and especially distribution systems controlled by retailers.

7. Transport costs are generally around 40% of total distribution costs (which include warehousing, handling, stock financing, administration, order processing and outer packaging), and thus amounted to nearly £5bn in 1982. 99% of this is in road transport.

8. The other main trend is the move from own-account operation to contract distribution. In one respect it is a natural response to the decline of suppliers' own distribution activities and the consequent move into "third party" distribution by supplier's erstwhile distribution divisions. Some of which, having become separate companies, have floated off: one thinks of SPD, formerly of Unilever, now integrated into National Freight Corporation.

9. How does this come about? The decline of own-account operation, perhaps more than anything, reflects the power of information technology to substitute 'control by information' for 'control by doing it yourself'. Systems used by Sainsbury's for branch ordering, depot replenishment, stock control and branch order picking and assembly, mean that effective control of activity in one of our contract depots is exactly the same as that in a Sainsbury-owned depot. The performance of the distribution network is therefore transparent to ownership of the distribution resource.

10. The significance of this is that it creates a 'market' for fully integrated and professionalised distribution activities. While a free market has existed for road transport since the deregulation - the abolition of quantity licencing - in the 1968 Transport Act, a

INDUSTRIAL VIEW

market for the total distribution service has been more recent. The significance of the existence of a market is that it makes reaction and adjustment to change of both the client and the provider that much easier and quicker. In particular, it means that a particular retailer's depots can more readily be relocated and/or expanded as a result of, for instance, major changes in the road network.

ROLE OF TRANSPORT AND THE ROAD SYSTEM

11. The significance of the move towards retail-controlled distribution is that it involves using transport in a different way. As explained above the economic advantage arises because of the consolidation that takes place at the retail-controlled distribution centre. In food, for instance, goods from a very wide range of suppliers are received, stored, picked, and assembled and delivered to branches to meet standardised order cycles and delivery schedules.

12. Commodities held under different temperature conditions - chilled, frozen, ambient and produce - are assembled and delivered in mixed loads, according to rules, during the course of a trading day. The objectives are to meet the branch requirements for perishable goods early in the day, and to achieve optimum transport utilisation.

13. The effect of consolidation is that delivery to branches is mostly a trunking operation - single load, single drop. The contrast with the typical supplier-controlled operation, basically multi-drop, is striking. The principal impact is on the amount of time that can be spent driving from the depot to the point or area of delivery. In trunking, the time spend unloading can be as little as 10% of the driver's day, so the driving time between depot and branch can be up to 45% of his day in each direction.

14. By contrast, a typical multi-drop could involve as much as 30% of the driver's day driving between drop locations, 40% of his time waiting to deliver and unloading, and only 30% 'stem' driving - 15% in each direction - between depot and delivery area. The range reachable from one depot in the former case could be up to three times as much as in the latter.

15. The determining factor in depot location and the relationship of depot to retail outlets to be served is the driver's maximum working hours ("duty time") and the maximum driving hours. These are laid down in EEC regulations; new revised regulations are due to

come in next September. While the regulations offer flexibility within a week and within longer periods as to exactly how the weekly and fortnightly maximum driving hours can be allocated, the effect is broadly one of setting a practical daily maximum around which transport schedules can be built.

16. A 'patch' of branches served by a depot would therefore normally be limited by the time/distance within which a driver can go out and return within the day. With the trunking pattern of work dominant in retail-controlled distribution, the primary road network is crucial not only in depot location, but in the viability of certain more far flung retail sites.

17. Certain distribution operations in retail involve a 'night out', but this is usually disproportionately expensive because of the driver's accommodation and subsistence costs, and the lower vehicle utilisation unless change-over is achieved. It may be impracticable with perishable goods in any case.

18. Therefore the move towards retail-controlled distribution from supplier-controlled distribution is a move towards trunking and away from local deliveries, with a consequent utilisation of larger and maximum gvw vehicles, and most importantly a greater use of and dependence on the primary route network.

19. In locational terms, supplier-controlled or wholesaler distribution into retail outlets tends to have more depots in or near towns, so that the drops can be maximised with relatively small 'stem' driving time. Retailer-controlled distribution, based on full trunkers into a more limited number of branches, will tend to have fewer depots, located on or near the primary road network, probably away from large towns, to take maximum advantage of the much longer driving time available in the shift.

20. The difference between supplier-controlled and retail-controlled distribution can be illustrated dramatically with an illustrative 'model'. Suppose there is a region in which 5 retailers each owns 50 supermarkets - total 250 - all of which sell goods provided by 25 different suppliers.

    The Appendix describes the model in detail. The key point is that the transport and warehousing efficiency of the retail-controlled network is much higher. The characteristic of the retail-controlled operation is:

INDUSTRIAL VIEW

- small number of large depots
- use of large vehicles
- single drop, trunking
- large geographic region covered from each depot.

This type of operation is typically planned around, and heavily dependent on the primary route network.

The characteristic of the supplier-controlled operation is

- large number of small depots
- use of smaller vehicles
- multi-drop, local delivery
- small area covered from each depot

21. Location of principal Sainsbury depots, for example, in relation to the supermarkets, illustrates the dependency on the primary route network. The most dense location of branches is in Greater London and the Home Counties, including the South coast and East Anglia. The next most dense area is the Midlands, and then the region to the north, as far as York and Lancaster, and South Wales and the south-west. Three principal depots - at Buntingford (on A10), Basingstoke (M3) and Charlton in SE London - serve the first area; Droitwich (M5/M42), Middleton, on the north side of Manchester (for M62) and Yate (north-east of Bristol - M4/M5) serve the others. These mostly carry a full range of chilled goods, produce, frozen and ambient grocery and non-foods, and each has a patch whose transport geography reflects the delivery disciplines of "overnight" perishables and produce.

22. Fifteen other depots carry a generally complementary range of goods, on similar and different order cycles and lead times; it is significant that seven of these are very near the M25, and a further three within easy striking distance. Two of these depots carry out a national or near-national distribution of slow-moving goods.

IMPACT AND VALUE OF PRIMARY ROUTE NETWORK IMPROVEMENTS

23. The impact of network improvements is to enlarge geographically the effective patch of a depot, given the practical constraints on driving time and duty time outlined above. The value of this can be taken in one of two ways, or in combination.

i) the feasibility of serving branches located further away, which opens up new market potential for the company.

ii) the ability to reduce the number of depots serving a total geographical territory.

24. It is fascinating to trace the historical growth of Sainsbury branches from the early days, in the late 19th century and the first quarter of this. When the company moved its depot to the Blackfriars site (where the head office still is today) before the turn of the century, branches were being established along the main radial routes out of London, within the catchment of horse drawn transport, and subsequently of motor lorries. The original concentration of branches on, for instance, the main Brighton Road can still be detected in the current supermarket locations. For a while, branches in the Home Counties and on the south coast were served by the pre-eminent primary network of the nineteenth and early twentieth century - the railways. The logistics of handling fresh product, assembled for specific branches, on and off the railway was challenging, to say the least.

25. Today the primary network permits the servicing of a new branch at Plymouth, an existing and a new branch at Exeter, existing branches at Swansea and Bournemouth - all from a principal depot near Bristol. Lincoln, Lancaster and Washington New Town SavaCentre are serviced from Middleton; Great Yarmouth and Corby from Buntingford, and Dover and Eastbourne from Charlton. The operational requirement consists of receiving into depot produce and perishable goods up till 2000, receiving in mid-evening the branch orders placed earlier in the day, and commencing assembly after 2200, with first vehicles generally away after 0300. Most branches start to receive 0630, 0700 or 0730; a few earlier than that.

26. A key feature of the road network which governs the definition and scheduling of work on the depot's 'patch' is the predictability of the journey times. Since the costs, not to say the inconvenience as well, of a driver going 'out of time' because of journey delays are so considerable, any experience of unreliability of journey times has to be reflected in conservatism of estimating. Inevitably, such conservatism will reduce the transport utilisation.

27. Reliability of journey times (outside the conurbations) is related to

- adequate road capacity in relation to demand

- the by-passing of bottlenecks at critical and unavoidable locations

- the management of essential road maintenance to minimise cumulative delays.

28. Examination of the network of transport operations from Sainsbury depots shows a number of key features. These derive from the location of depots as described in para 25 above, and the spread of branches, most densely in the Greater London conurbation, and at lower densities in the Midlands and even lower in the North, although JS is well represented in the freestanding towns in the Midlands and SE.

- the value of the 'figure of 8' Motorway network, and M5 to the south-west.

- the importance of the M25, for instance, for part-circumferential journeys to access London branches from radials.

- the inadequacy of many cross-regional routes (with notable exceptions of A34 and A45) - the 'secondary' primary network.

29. It is encouraging to see a recognition by DTp in the 1985 Roads Review of the importance of the 'secondary' primary network, most of which were highlighted in the 'Fabric of the Nation'. The requirement is not primarily one of much increased capacity - except at known bottlenecks - it is the ability to achieve predictable speeds at a consistently high level (not necessarily always at the legal maximum). With predicted vehicle flows that are much less than those which justify motorways, value for money criteria understandably influence the characteristic, particularly whether dual 2-lane or single carriageway.

30. While speed characteristics of newly built single carriageways with good sightlines and sympathetic alignment are certainly beneficial, the performance characteristics of many HGVs do not necessarily lend themselves to making the best of the more 'opportunistic' driving that good single carriageway roads can offer. This reflects back on the predictability element of journey times, and the inability to schedule for the benefits of road improvements unless there is a high level of guarantee. The dual 2-lane road offers a degree of predictability that under a much wider range of

traffic conditions enables the benefits to be realised in planning and scheduling times.

31. It is arguable whether current methods of cost-benefit assessment fully account for the benefits of network improvements, in view of the structural changes in distriubtion logistics can bring, as described in para 23 above. Time savings to commercial vehicles are typically evaluated at a rate which corresponds to the times costs of operating the vehicle, inlcuding the driver. Conventional economic thinking would attribute any consequential "business" benefits of vehicle time savings as being equal to (or less than) the original benefits, on the grounds that if there were greater consequential benefits some other way would have been found to realise them. This is logically attractive, but it presumes that resources are infinitely adjustable and substitutable according to certain relationships. The reality of transport operation is that there are discontinuities - most particularly, as explained, arising from the drivers' hours limitations.

32. Let us consider, within for example food retailing, the effect of a set of primary hypothetical network improvements such that a given total territory could be served with 5 depots in place of 6, with the same resulting total travel time. The benefit can be taken either as

    i) reduction in the number of depots. If the total travel time remains the same, the average distance travelled increases. Against this are the savings at average cost of a complete depot, offset by the increase in volume through the remaining five depots at marginal cost. Finally, there is a stockholding saving, since with 16% more branches at each depot the weeks' cover to maintain a given level of service can be less.

    ii) time savings, without change in the distribution network. This is the evaluation included in the cost benefit road assessment.

33. On reasonable estimates of the elements of distribution and stockholding cost, it can be shown that the savings from reducing the number of depots can exceed the straight value of time savings by 30-50%. While it is appreciated that cost benefit tools may do little more than provide a rationing and prioritising device for road schemes and policies within an externally determined financial ceiling on road expenditure, it is valuable and important that the "business potential"

INDUSTRIAL VIEW

unlocked by network improvements should be better understood. It might indeed throw further light on the single versus dual 2-lane carriageway debate. This section does no more than illustrate the means by which this business potential can be unlocked in retail physical distribution (where controlled by the retailer).

CONCLUSIONS

34. Changes in physical distribution in retailing are making the whole operation more dependent on - and more able to take advantage of - the primary route network. The continuing move from supplier-controlled distribution towards retailer-controlled distribution substantially changes the economics of the transport element; the move away from own-account to contract distribution creates more of a 'market' for distribution services, and this enables distribution networks to react and adjust more readily as changes in the road network come along.

35. The governing elements in efficient transport scheduling are the drivers hours' limitations imposed by EEC regulation. A key feature of any road network improvement is what it does for predictability of journey times, as well as for average journey times; the dual 2-lane carriageway has particular value because it does not demand the degree of 'opportunistic' driving that single carriageway improvements do, and which it is difficult for (laden) commercial vehicles to take full advantage of. Benefits to commercial vehicles of road improvements, calculated as straight time savings, will tend to understate the true "business potential". It can be shown that, in a typical operation of retail distribution of food, the benefits from restructuring the distribution and depot network could exceed the straight time savings by 35-50%. It is important that the "business potential" released in this and other industries by network improvements are better understood.

Ref 1: National Economic Development Office, Distributive Trades EDC. "Factors Affecting the Cost of Physical Distribution to the Retail Trade". November 1985 unpublished.

Appendix

Model comparison of supplier-controlled distribution networks with retail-controlled distribution.

A region contains 5 retailers, each retailer having 50 supermarkets which could be served from one depot. Each supermarket sells 1000 cases/day in a five day week; goods come from 25 suppliers in equal quantities; 1 roll cage holds 20 cases; a large vehicle holds 50 roll cages.

Model A: Supplier-controlled distribution

The demand from each of the 250 supermarkets on each of the 25 suppliers is 40 cases/day, which is 2 roll cages/day. This would involve 25 drops/day from one large vehicle which, depending on the geography, is unlikely to be practical. If all the goods were short life perishables, the daily delivery requirement can only be met with a larger number of smaller delivery vehicles - for instance a 20 roll cage capacity vehicle working 10 drops.

Non-perishable commodities would be on a less frequent cycle - say twice/week, so each large vehicle would make 10 drops of 5 roll cages each; the retailer would have to hold more stock.

If half the suppliers were perishable, and half non-perishable, then the transport requirement for each supplier, and in total, would be

	each supplier	total
12.5 perishable suppliers	25 small vehicles	313 small vehicles
12.5 non-perishable suppliers	10 large vehicles	125 large vehicles

The 'stem' time available for a 10 drop journey is at best a third of that available to the retailers' own single drop transport under Model B, so the territory that each supplier's depot could cover is at best one third. If the region is such that each retailer could service its own branches from one depot, then in Model A each supplier would need 3 depots. The average depot size for each perishable would be 8.33 vehicles, and for each non-perishable supplier 3.33 vehicles. In practice, some trade-off between number of drops/stem time/number of depots/number of vehicles would be made; for instance, the non-perishable supplier might manage with 2 depots if the number of drops reduced to 5, and twice the number of 25 roll cage vehicles were used. The average number of vehicles per depot would then be 10.

INDUSTRIAL VIEW

Model B: Retail-controlled distribution

For each of 5 retailers, one depot serves 50 branches. All
suppliers deliver into each of 5 depots (compared with
delivery into 2 or 3 of their own depots under Model A).
Each branch receives 1,000 cases/day, in 50 roll cages,
equal to one vehicle load. The furthest branches in the
region take a whole driver's shift to complete the return
journey; for nearer branches two or in some cases three
return journeys can be completed within a shift. If the
average number of return journeys per shift achievable is
(say) 1.67, then the transport requirement is:

each retailer depot	total
30 large vehicles	150 large vehicles

This requirement is met from a total of 5 depots, with 30
vehicles per depot. Both models assume 1 driver's shift
per vehicle; if delivery windows allow, this could be
assumed to increase, although there is more scope under
Model B because the 'stem' time out on the first journey
and back on the last can be outside branch receiving hours.

Comparison:

	no of vehicles	no of depots	average depot throughput
Model A:			
perishable	313 small	37.5	3,300 cases/day
non-perish	125 large	37.5 or 25	3,300 or 5,000 day
Model B:	150 large	5	50,000 cases/day

2. Road links for chemical exports

D. J. BRICKNELL, BSc, PhD, Company Distribution Adviser, Imperial Chemical Industries plc

SUMMARY. This paper uses an analysis of the changing structure and market emphasis of Imperial Chemical Industries Plc (ICI) as a model for the UK Chemical Industry and demontrates that a comprehensive and adaptable road infrastructure is essential for the continuing success of this major exporting industry.

INTRODUCTION
1. ICI can be accurately described as a multinational chemical company based in Europe with its headquarters in the UK. In the period from its formation in 1926 to perhaps 1960 it would have been properly described as a UK chemical company with strong exporting links with the Empire and Commonwealth. The difference highlights the way the Company has adapted its structure and businesses to suit the changing world.
2. A multinational chemical company survives and thrives today by ensuring diversity in both its products and the geographical location of its operations, by a rapid and flexible response to the disparate needs of its customers and by being able to adapt and evolve its businesses in step with the rapidly changing political and social patterns in the world markets. It is well known that products have a life cycle and a successful company must manage its portfolio of products to maintain a good distribution of new specialty products for which the market needs to be developed, of mature products for which the markets are established, but for which new uses can be found, and old absolescent products for which the uses and markets are declining. The different commercial needs of these products are reflected in a wide range of transport and distribution systems.
3. Chemical companies typically spend about 10% of the value of their sales on the distribution of their products, but the proportion may be as low as 5% for certain high value products such as dyestuffs or pharmaceuticals and as high as 25% for bulk commodities like caustic soda or chemical fertilisers. The major part of this expenditure (typically about 50%) on distribution is accounted for by

INDUSTRIAL VIEW

the cost of transporting the products from the production site, sometimes via an intermediate warehouse, to the customer. Whilst some use is made of all of the conventional means of transport by far the most important is road; this is true not only for sales within the UK but also for the majority of exports to "short-sea" West European markets.
4. Efficient distribution systems are as essential for the success of a modern chemical company as are technically advanced and energy efficient production plants and sophisticated products. In today's competitive environment the difference between profit and loss on business in chemicals can often equate to the skill and imagination shown in the choice of distribution system. Chemical companies now recognise the need for fully integrated distribution processes in which package size and design are optimised for efficient use of transport and storage, the distribution method is optimised to minimise product inventory and therefore working capital required, the use of expensive intermediate storage and transhipment is avoided and in which effective use is made of multi-modal transport. The end result must be a distribution process which reflects the service needs of the customer at a cost which the particular business can profitably bear.
5. Chemical companies recognise the importance of efficient and safe distribution of their products and deploy senior management effort to the control of this function. They recognise the vital role of road transport in the distribution of chemical products and in consequence have a strong interest in the creation, maintenance and management of the road infrastructure of the countries in which they operate.

THE ICI EXPERIENCE
6. The Company is in many ways a microcosm of the UK chemical industry; ICI accounts for about 25% of the UK industry's domestic sales and a similar proportion of the UK's chemical exports; ICI is currently exporting chemical products at an annual rate of £2.5bn a year.
7. The ICI Group comprising over 300 subsidiary and related companies ranks first in the world on financial performance and fourth or fifth in the world in size depending upon the measure adopted. It has manufacturing facilities in over 40 countries and over 200 manufacturing sites world wide. The Company's product range extends from agricultural, industrial and speciality chemicals, fibres, petrochemicals and plastics, paints, dyestuffs to industrial explosives and pharmaceuticals; the diversity of the product range is probably unequalled by any other major chemical company. ICI also has significant interests in oil. The Company's very wide product and geographical spread gives it an excellent basis for success provided that the marketing and distribution of the diverse products from the many

production sites to the multitude of customers can be managed cost effectively; hence the emphasis placed by ICI on sound distribution. In the ICI context "sound" distribution implies not only the technical competence to be aware of and apply the best available technology (unit loads, containerised movements, intermediate bulk containers, 38 tonne gvw vehicles etc), but also the ability to adapt to changes in technology and the availability of services and to provide a flexible response to customers' needs (delivery by road today and rail tomorrow).

Product and Geographical Diversity
8. The relative importance of the Company's business areas and markets world wide are shown in Tables (1) and (2). Comparison of the figures for 1960 and 1970 with the present day figures for 1984 (all inflated to 1985 values) highlights the enormous changes which have taken place in the world markets over this time scale and the response these changes have prompted from ICI.
9. Table (1) shows the way the export business of the Company has changed since 1960. Firstly the growth in the importance of exports as a proportion of the Company's sales; an increase from 17% of total sales in 1960 to 24% in 1984. Secondly, and perhaps more importantly, the growth of the UK as the main export platform for the Company. Whilst in 1960 exports comprised 24% of the total sales ex-UK this proportion had grown to nearly 50% by 1984; put another way, about half of the chemical products made by ICI in the UK are now exported. Finally, Table (1) illustrates the declining importance of the UK domestic market; whereas in 1960 the UK accounted for 72% of ICI Group sales this proportion had diminished to 51% by 1984.
10. Table (2) illustrates the marked change in export destination for the Company's products over the period 1960-84. In 1960 whilst Continental Western Europe (CWE) accounted for 36% of the total exports by value the commonwealth countries matched this with slightly over 37%. By 1984 not only had the total quantity of exports substantially increased but the proportions to these markets were quite different; CWE then accounted for 52% of the total whereas the commonwealth countries share had declined to less than 30%. The impact of membership of the EEC and the decline of the economic importance of the Commonwealth markets to the UK are clearly demonstrated.
11. The relevance of these major changes in market patterns for the planning of the UK's road infrastructure lies in the different routes and methods of distribution which the new market destinations and customer requirements have caused to be developed over the 1960-84 period. For example, the invention and widespread introduction of the ISO freight container has led to the funnelling of goods through the limited number of ports equipped with the necessary storage

INDUSTRIAL VIEW

TABLE 1. ICI's Changing Shape

1985 £m

	1960	1970	1984
ICI Group Sales	4103	5985	9121
Sales ex UK	3180	4683	4613
Exports ex UK	713	1252	2183
1) Exports as % of UK Sales	24	27	47
2) Exports as % of Total Sales	17	21	24
3) Sales ex UK as % of Group Total	72	78	51

TABLE 2. ICI's Changing Export Markets

1985 £m

	1960	1970	1984
Export Sales to:			
EEC	102	254	950
Total CWE	255	584	1136
CWE as % Total Exports	36	47	52
EEC as % Total Exports	14	20	44
Americas	93	173	316
Americas % of Total Exports	13	14	14
Commonwealth	266	361	538
Commonwealth % of Total Exports	37	29	25
Total Exports ex UK	713	1252	2183

and handling equipment for containers. These movements also incidentally provided the base-load for British Rail's "Freightliner" service although many containers still move to ports by road.

12. Until the 1960's the bulk of the UK's chemical and other exports were destined for deep-sea (commonwealth) markets, and the ports of Liverpool, Glasgow, Bristol, Southampton, Hull and London thrived. Once the export emphasis switched towards CWE these ports entered into a period of pronounced decline. This was caused primarily by an unavoidable geographical disadvantage for the West Coast ports but made much worse by an inexcusable combination of management ineptitude and work force intransigence. The resulting poor standards of service, frequent disruption and uncompetitively high costs ensured that customers went elsewhere. The results of this protracted warfare in the UK's major ports are still being felt and are relevant to the future consideration of road infrastructure.

13. The increasing demand for efficient services between the UK and CWE has led to the rapid evolution of roll-on/roll-off (ro-ro) and lift-on/lift-off (lo-lo) services using large (38te, 12m) articulated trailer vehicles of many different types and very sophisticated ro-ro ships which provide a rapid and economical crossing of the short stretch of water.

14. The promise of a fixed link across the Channel is greeted with enthusiasm by ICI and the chemical industry for the increased flexibility in distribution operations which such a link could provide; but here too the road infrastructure to serve such a link must be created simultanesouly with the link itself. The mistake of failing to provide an adequate Thames crossing to link the M25 motorway at Dartford must not be allowed to happen again at Dover or Folkestone.

15. The rapid growth of ro-ro and lo-lo services from the UK to CWE has required that the UK ports on the South and Eastern seaboards of the UK should display the adaptability and flexibility so obviously lacking in their deep-sea counterparts. The chemical industry welcomes the successful enterpreneurial response which has been forthcoming and the ports of Felixstowe, Dover, Folkestone, Immingham and numerous smaller ones have deservedly thrived. All that now remains to complete the success story is the provision of an adequate road infrastructure to enable the UK's chemical exports to travel to these ports speedily and economically.

Distribution in ICI

16. The distribution function in ICI embraces a wide range of inter-related activities; these include packaging and labelling of the product, storage and warehousing, transport to the customer and all of the computerised business systems for order processing, documentation preparation and

customer invoicing. Production sheduling, inventory management and raw material purchasing are also sometimes included in the manager's portfolio. The distribution manager in ICI has full control of all of the processes necessary to deliver the required goods to the customer in a sound state and on time - provided of course that the production plant has actually produced an adequate supply of the right product! ICI spends about 8% of its net sales income on distribution, currently this amounts to about £800m per annum. About half of the distribution expenditure is accounted for by the cost of transport services bought in (90%) and operated in-house by ICI (10%). The breakdown of the transport expenditure between the modes and the changes in the relative use made of each over the period 1960-84 are shown for ICI UK in Tables (3) and (4).

17. The figures demonstrate the marked decline in the importance of rail transport and commensurate increase in the use of road transport for the movement of chemical products over the 24 year period. The Company expects that the major decline in the use of rail transport has now come to an end; the residue of the Company's products (about 10% by tonnage) now carried by rail is expected to be maintained to provide a strategic and commercially effective alternative to road transport and because of the strenuous efforts being made by the transport service industry to develop economical multi-modal transport systems involving rail movement for part of the journey.

18. The sea mode proportion of freight expenditure is shown to have stayed substantially constant at about 50% over the period, but this overall figure hides substantial changes in the proportion of liner freight, chartered vessels and of short sea ro-ro expenditure. A substantial part of the 1984 exports start their journey from the UK by road or rail.

19. The dominant role of road transport in the movement of chemical products both within the UK and for export seems unlikely to be challenged in the future. It follows therefore that the chemical industry fully supports the provision of the best possible road infrastructure for the UK. Only then will the road transport service industry be able to provide the required standard of distribution service at an acceptable financial and environmental cost.

THE REQUIREMENTS FOR A ROAD INFRASTRUCTURE
20. It is self evident to any one using the road network in CWE that the provision of an adequate road infrastructure in the UK has lagged a long way behind that of our European competitors; even in 1985 the basic infrastructure is incomplete. Where it is complete it is all too often showing signs of either premature failure or of having inadequate capacity for even todays traffic density. The UK's road building programme appears so far to have achieved the construction of roads of inadequate size and strength to

TABLE 3. ICI UK - Use of Transport Modes

	1960	1974	1980
1) Road	15	42	47
2) Rail	33	8	5
3) Sea	52	50	46
4) Air	<1	<1	2
	100%	100%	100%

TABLE 4. ICI UK Transport Expenditure

1985 £m

	1960	1974	1984
Road	35	123	115
Rail	80	20	12
Sea	125	143	113
Air	<1	1	5
Total	240	286	244
Tonnage	20×10^6	20×10^6	20×10^6

destinations which no longer reflect the needs of the UK's exporting industries. The chemical industry sees the provision of a complete, soundly constructed and adequately sized road network as fundamental to the continuing success of its export business and for the safe and environmentally acceptable operation of its business within the UK.

21. The parallels with the chemical industry are marked. In the 1960s industry, including ICI, were much wedded to the idea that by the use of sophisticated mathematical techniques to analyse historical data made possible by the use of the most advanced available computers accurate meaningful forecasts could be made about the future! Many large production plants were constructed on the basis of such forecasts and are currently lying idle or under utilised because the forecasts turned out to be wrong. This does not of course mean that trying to forecast the future is a futile exercise, just that all forecasts should be viewed with scepticism and used to support and not replace professional business judgement. This message would seem to be as applicable to the planning and construction of road infrastructure as to the planning and construction of chemical plants.

22. The chemical industry has tackled the problem of operating profitably in an unpredictable market environment by a variety of methods; ICI's approach has been to create a smaller, simpler more responsive organisation; this has been achieved by employing fewer people in fewer levels of management and by devolving much decision making to those running individual businesses. The aim has been to enable the organisation to respond rapidly and effectively to unpredictable external events. In planning future action businesses consider the most likely of a range of possible business environments and ensure that even if the worst should happen a credible plan for survival exists. If the external economic environment turns out to be unexpectedly favourable then the business must be equally able to capitalise on this outcome. The aim is flexibility and adaptability in coping with uncertainty.

23. The author believes that a similar approach would have benefits for the planning and provision of road infrastructure. It is futile to try to predict which goods or people will move by which method or mode of transport to which destinations. It must surely be better to create an infrastructure which preserves industry's options to respond flexibly to future demands, whatever they may be without suffering the constraint on business performance of an inadequate road infrastructure.

CONCLUSION

24. The chemical industry needs a widely spread, adequately sized road network linking all major conurbations, industrial areas, ports - particularly the ro-ro and lo-lo

ports on the Southern and Eastern coasts of the UK – and the rail-road bimodal interchange terminals now being created.
25. Forecasts of traffic flows to and from UK ports should be viewed with considerable scepticism and roads should be of a size and strength to accommodate the largest traffic flows and the heaviest vehicles which can sensibly be envisaged.
26. Additional emphasis should be placed on completing the obvious gaps in the UK's road infrastructure, for example, the provision of good access to the ports of Felixstowe and Harwich. On rebuilding and strengthening those existing roads which are proving to be inadequately constructed and on the addition of capacity to those roads already over-loaded and to those clearly about to become over-loaded in the immediate future.
27. The chemical industry's position as the UK's top exporter of manufactured goods will continue to depend upon economical and safe distribution; efficient road transport is the back-bone of the economic distribution of chemical products. A comprehensive and sound UK road infrastructure is not a luxury which society cannot afford but a necessity upon which the continuing success of the industry depends.

Discussion on Papers 1–2

DR QUARMBY
In introducing the discussion session, I would particularly draw attention to the debate about the dual carriageways versus high standard single carriageway roads. I would argue that taking full advantage of high standard single carriageway roads requires a degree of opportunistic driving which HGVs, with their low acceleration characteristics, may just not be able to achieve. By contrast, it is the predictability of journey times which dual carriageway roads allow HGVs to achieve that gives the real value, as journey planning requires a degree of certainty - the 90 percentile journey time. While good average speeds may be achieved on standard single carriageway roads, the lower predictability makes them of less value to HGVs. My question is this - does the choice of either a dual two-lane or a high standard single carriageway road reflect these factors?

MR W. BRADSHAW, British Railways Board
With reference to Papers 1 and 2, I would like to support the view that, in the planning of the primary route network, the requirements of industry should take precedence. As a former member of SACTRA, I am convinced that COBA is useful as a tool only for ranking and fine tuning.

My experience on Merseyside in the 1970s has led me to believe that motorway construction is not the catalyst for industrial regeneration that some people think it to be. On the other hand, centres with growth potential, such as the East Anglian Ports, obviously need good roads.

When new roads are constructed, prudent provision should be made for increased traffic flow. In this context, it is relevant to recall that when Brunel built his railway from Paddington, sufficient land was acquired for up to six tracks, although only two were laid. It is apparent that DCF techniques are inadequate to express the costs of widening a congested road at a later date.

Lastly, I would ask whether any experience has emerged of the effect on HGV speeds of the adoption of Highway Link Design Criteria (TD9/81).

INDUSTRIAL VIEW

MR G. R. CAMERON, Warwickshire County Council, Department of Planning and Transportation

Single carriageway road capacities should take into account the number of HGVs using individual roads. The M1/A1 Link Road to East Anglia was increased in standard from a three-lane to a dual two-lane road on account of the large number of HGVs (particularly chemical tankers) passing along it. Also, crawler lanes should be provided on hill sections of three-lane roads where lorry flows are high.

There is a need to fill lorries returning empty from jobs in order to reduce total lorry traffic.

The primary route network is resulting in pressure from many quarters for the creation of shopping parks on the fringes of towns; such pressure may be difficult to resist.

DR D. COOMBE, Halcrow Fox and Associates
Contribution 1

As someone who was partly responsible for some of the preliminary research, I would like to make the following point in response to Dr Quarmby's question about Highway Link Design.

After research had been carried out on a wide range of road types - from tortuous roads to modern 10 m single carriageways - relationships between heavy vehicle speeds and road geometry were established. These acknowledged relationships were subsequently reflected in the Highway Link Design. However, as far as I am aware, no check has yet been made on the relationships that exist in practice on the resulting new single carriageways.

Contribution 2

As part of a recent review of urban roads' appraisal methods for the DTp by Halcrow Fox and Associates, development impacts caused by new or improved roads were considered. Economies of scale are likely, but are very difficult to identify, as frequently a myriad of microscopic effects result. The directly estimated economic benefits (time savings, vehicle operating cost savings and so on) omit very little, and these direct effects are transmitted into other parts of the economy; therefore, we must be aware of the danger of double-counting. Roads often have two ends, making the redistributional effects difficult to predict in some cases. However, it is important to try and identify them, and to include the revised planning inputs in the traffic forecasting process. The resulting direct economic benefits should thereby omit very little from the total economic benefit of road schemes.

MR T. PUMFFREY, Department of Transport

It has not yet been possible to assess the effect of Highway Link Design on road performance, as so few of these roads have

DISCUSSION

yet been completed and opened. Most single carriageway roads opened of late have been designed according to the old criteria. The latter allow a free-flowing and curving alignment; this tends to encourage speed which, in turn, gives rise to a bad accident record. Highway Link Design roads — with straighter alignments, short dualled sections and crawler lanes, which should give lorries and other vehicles much more opportunity to overtake — are expected to achieve a better safety record. Such a road design could be termed 'opportunistic'.

With regard to design speed, enough weight is perhaps not being given to heavy lorries using the 85 per cent approach.

Value for money must be accepted in respect of dual two-lane as against single carriageway roads. The question is whether, with a finite amount of money available, the Authors would rather have one dual two-lane bypass, and another scheme delayed, or two well-designed single carriageway bypasses.

DR S. T. ATKINS, Greater London Council
In response to a comment by Mr Bricknell, Author of Paper 2, concerning the GLC's lorry ban, I would like to point out that it is the virtual completion of the M25 that has allowed the development of the GLC's modest restriction. I am pleased to hear that Mr Bricknell's organisation would not route through lorries by way of inner London. However, from his own admission to having to give instruction to his drivers to avoid central York, such persuasion is clearly necessary as one cannot rely on the initiative of individual drivers to comply with environmentally beneficial route guidance. With reference to the performance of the lorry ban, it is my understanding, from limited studies undertaken in the initial weeks following the introduction of the measure, that the early morning flows of the relevant lorries have been reduced by 30-50 per cent.

MR V. E. JONES, County Surveyor and Bridgemaster, Hereford and Worcester County Council
I would like to draw attention to the increasing use of the road mode for the movement of ICI products (Tables 3 and 4 of Paper 2). It could be deduced from these tables that the tonnage of products moved had increased from 3 million tonnes in 1960 to 9.4 million tonnes in 1980, a growth of over 200 per cent. However, expenditure on improvements and additions to the primary route network has failed, by a wide margin, to keep up with this growth. I would urge that greater consideration should be given to the relationship between industrial locations and road planning. Some of the problems which have already been identified on the M25 can be attributed to the traffic generated by the industry which has established itself on the western edge of London contiguous to the motorway, and around the airports at Gatwick and Heathrow.

3. An overall review of the national network

P. E. BUTLER, Head of Local Roads Division, Department of Transport

SYNOPSIS
The primary route network (PRN) is the joint responsibility of central and local government. It operates within the framework of overall policy on roads and allocation of resources. The Department of Transport gives priority to new construction and structural renewal on its half of the PRN while not neglecting the need for normal maintenance. Local highway authorities are encouraged to give suitable priority to their roads of more than local importance, including the PRN.

INTRODUCTION
1. The PRN epitomises the partnership between local and central Government in providing the country's main road system and illustrates a major focus of central Government attention and resources. The network of primary routes signed in green, designated 20 years ago to guide through traffic on longer journeys between places of traffic importance, today consists of national and local principal roads in virtually equal proportion. A glance at the White Paper "Policy for Roads in England: 1983" (Cmnd 9059) and the Public Expenditure White Paper 1985 (Cmnd 9428) shows emphasis on the provision of primary routes, the reduction of delays and transport costs, the improvement of urban roads and the relief of communities from heavy through traffic.
2. There are some 264,000 kilometres of roads in England as at 1 April 1984.

Table 1. Roads in England 1984

Classification	Kms	%
National Motorways	2,360	1
National Roads	7,850	3
Principal Roads	24,710	9
Remainder	229,040	87

CENTRAL AND LOCAL GOVERNMENT POLICIES

3. The PRN consists of some 18,000 kilometres of this total, all the national (trunk) roads and 40% of all local principal roads. It carries over 45% of all motor traffic on local principal roads and over 50% of heavy goods vehicles. National roads carry some 40% of all the traffic on the PRN but about 60% of heavy goods vehicles kilometres are driven on them.

Table 2. All Motor Vehicle Traffic (million vehicle kms) 1983 (latest available)

National	Principal Primary	Principal Non-Primary	Total	% of Primary Traffic on National Roads	Primary Traffic as % of Principal Roads
29,500	41,500	47,000	118,000	42	47

Table 3. Heavy Goods Vehicles (million vehicle kms) 1983

| 5,075 | 3,150 | 2,825 | 11,050 | 62 | 53 |

4. The PRN operates within the framework of overall policy on roads. In what follows, therefore, what holds true in the general sense also applies in equal measure to the PRN.

Highway Responsibilities

5. The Secretary of State for Transport, while directly responsible as highway authority for the national motorways and national roads in England, also has a more general responsibility. He is concerned that there is an appropriate road system for the country as a whole in line with government policies for the economy, industry and commerce, that the entire system operates safely, effectively and efficiently with due regard to the environment, and that allowance is made for the higher road traffic growth currently forecast.

6. The county councils, the Greater London Council and the London boroughs are the highway authorities for all other roads. On 1 April 1986, after the abolition of the Greater London Council and the metropolitan county councils, London boroughs and metropolitan district councils assume full highway responsibilities for the local roads in their aras.

Heavy Through Traffic

7. An overriding objective of the Government is to make due allowance for the general flow of through traffic and in particular for heavy lorries. The constant search for the most economic movement of goods in heavy vehicles places a great strain on the existing roads, conflicts with the quality of life for many people and with the environment in many places and calls for increased recognition in planning for future provision. The policies of the Department of Transport are focussed sharply on improving the system of through

routes, on the provision of bypasses and on the standards of bridges.

8. The European Economic Community is seeking uniform acceptance of heavy lorries and Parliament has already agreed the introduction of 38-tonne vehicles. The Community now wishes to introduce 40-tonne lorries, but the United Kingdom has secured an indefinite derogation in order to allow an assessment to be made of what further work on roads and bridges would be required if heavier lorries were allowed on the country's road system. The Community has to make a decision by 28 February 1987 on a proposal for a review procedure for the derogation.

Financial Resources for Transport

9. Financial resources available for transport must be set in the context of the Government's overall financial strategy:

(a) continuing the drive against inflation;
(b) helping create the conditions for more jobs;
(c) controlling public expenditure as the main-stay of policy.

10. These priorities form an integral part of the medium term financial strategy. Continued success requires steadily reducing public borrowing - broadly the gap between public spending and tax receipts - as a proportion of national income. At the same time the Government wish to provide a continuing reduction in the burden of direct taxation and a firm basis for economic growth. Cash-plans imply that public spending should remain constant in real terms for the current planning period, 1985/86 to 1989/90. Local authorities are responsible for about 25% of all public expenditure (about £34bn in 1984/85): the great part of this is current expenditure, which continues to exceed the Government's planned levels. For 1985/86 the Government spending provision increased by £950m to £25.5bn. Net provision for local authority capital expenditure for the same year was £2.2bn compared to £3.1bn in 1984/85. This reduction was partly to allow for increased receipts and partly to compensate for the £0.4bn overspend in 1983/84.

11. The Government's aims in deciding resources for road spending are:

(a) improving and maintaining the national road system;
(b) investing in local road improvement;
(c) maintaining local roads;
(d) promoting safety and environmental protection.

12. Transport programmes account for about 3½% of total public expenditure amounting to £4.7bn in 1985/86. Expenditure on roads accounts for about £2.5bn of this.

CENTRAL AND LOCAL GOVERNMENT POLICIES

Table 4. Transport Expenditure (mainly roads and bus/rail subsidies)

	Outturn £m			Planning £m	
	Final	Provisional	Estimate		
	1983/84	1984/85	1985/86	1986/87	1987/88
Central Government	1,629	1,905	2,318	2,400	2,380
Local Authorities	2,627	2,748	2,413	2,400	2,390
Total	4,256	4,653	4,731	4,800	4,770

Table 5. Roads Expenditure

	Outturn £m		Planning £m		
	1983/84	1984/85	1985/86	1986/87	1987/88
Central Government					
Capital	644^i	722^i	742^i)	870^{ii}	890^{ii}
Current	66	84	82)		
Total	710	806	824	870^{ii}	890^{ii}
Local Authorities					
Capital	457	460	502)	$1,610^{ii}$	$1,620^{ii}$
Current	877	(est)949	972)		
Total	1,334	1,409	1,474	$1,610^{ii}$	$1,620^{ii}$
Total All Roads	2,044	2,215	2,298	$2,480^{ii}$	$2,510^{ii}$

i. This figure includes expenditure on structural renewal, which is classified as capital. Road construction accounts for £514m in 1983/84, £568m in 1984/85 and £599m in 1985/86.
ii. Overall planning totals as at 1985 subject to revision. The figures for local authorities are notional and based on the assumption that roads spending accounts for the same proportion of provision as in 1985/86.

The figures show that, despite the pressures on public expenditure for 1985/86 and beyond, the Government recognise the high priority of transport in modern life.

Financing Transport Expenditure

13. The Department's spending on roads is financed directly from the Exchequer and voted by Parliament. Local authority current expenditure on road maintenance is financed by rates and rate support grant (RSG) block grant. Capital expenditure is financed by a mix of borrowing and rate-fund contributions which will vary between authorities. This spending is also supported by a Transport Supplementary Grant (TSG) (see paragraph 33). The balance of capital spending is supported

through that element of RSG which is paid in support of the assumed financing costs of new capital expenditure.

Rate Support Grant

14. Responsibility for administering RSG rests with the Secretary of State for the Environment. The starting point is the aggregate exchequer grant (AEG) available for local authority expenditure. This is determined by the Government in the light of its overall economic plans and in particular with regard to the control of public expenditure. For 1985/86 AEG (which comprises block grant, TSG and other specific grants) was about £11.8bn (some 49% of local authority relevant expenditure if authorities spend in line with PES provision). Block grant amounted to about £8.5bn.

NATIONAL MOTORWAYS AND NATIONAL ROADS

15. National motorways and national roads account for only 4% of the total length of roads in England. But they carry 30% of all traffic and 55% of freight traffic and, as has already been said, national roads make up about one half of the PRN. They receive about 60% of spending on roads-construction, amounting in 1985/86 to £599m with an additional £225m being spent on maintenance (Table 5). Since 1978/79 there has been a steady increase in capital spending overall with a marked emphasis on structural renewal which has trebled during the period. Of more than £3.8bn spent on the roads in this period some £1bn has gone on renewal and some £2.8bn on new construction.

Objectives

16. The Department's priorities are to reduce delays and transport costs to help economic recovery and growth, to improve the environment, especially by relieving communities of through traffic, to preserve the investment already made by adequate maintenance and to secure improved road safety.

Construction

17. *Establishing the Need* The Department adds schemes to its construction programme after a preliminary study of need and broad review of solutions and likely resources. A process of continuous review is summarised every 1 or 2 years in a White Paper (as in 1983) or a Roads Report (as in 1985) which formally adds schemes to the programme which are ripe for preparation. This continuing analysis is supplemented in two ways. There are occasional studies of need as that recently completed by the Department for the NEDC. There are longer term reviews of likely demand aimed not so much at immediate or early programme entry but rather at extending the planning horizon and ordering priorities more realistically within it. So the Department is beginning a comprehensive review of motorway capacity and of the remaining need for bypasses. Account must also be taken of current traffic forecasts. The revision of the National Road Traffic Forecast in 1984 was

followed by a thorough examination of the programme to ensure that schemes were of an appropriate standard.

18. **Meeting the Need**   The level of resources to be applied to construction rests on a judgement of competing needs. Between April 1979 and August 1985 184 major schemes have been completed, adding 852 kilometres of new road to the network, including 341 kilometres of motorways. Fifty bypasses and relief roads have been provided and 114 communities relieved of through traffic. Fifty-one schemes, including 110 kilometres of motorway and 21 bypasses and relief roads are under construction at a cost of £797m. In preparation are 286 new schemes with a total length of 1,760 kilometres, including 187 kilometres of motorway and 150 bypasses and relief roads. The programme will relieve 277 communities of through traffic at a cost of some £4bn. The Secretary of State's basic policy is that the design and supervision of major schemes should be carried out by firms of consulting engineers. Schemes valued at below £1m are allocated to local authorities as agents, as are a few major schemes where there are special reasons for doing so.

19. **Value for Money**   Value for money is a prime consideration. The economic case for road investment is assessed by cost benefit analysis program (COBA). Account is taken in this of savings in time, operating costs and accidents. These are translated into money-values for comparison with the cost of construction to establish net present value. Both design and timing of schemes need to be justified in this way. The total cost of planned scheme completion in 1984/85 to 1987/88 is estimated to produce total discounted user and safety benefits of about £2.5bn, ie £1.74 per £1 spent. This excludes substantial net environmental benefits which cannot satisfactorily be given a cash value and are not included in the COBA analysis. These are, of course, extremely important. They are dealt with in accordance with the Department's Manual of Environmental Appraisal and taken into account along with economics in the assessment framework.

20. **Pace of Construction**   The CBI, TUC and the Department recently jointly reviewed the possibility of accelerating certain schemes regarded by the CBI and TUC as the most important. They had identified some £3.68bn (at November 1982 prices) of such expenditure for the period 1985-1995. The current departmental programme sets out improvements worth £2.44bn (at November 1983 prices) on the same roads. In presenting the review Team's report to NEDC the Minister of State for Transport noted their comments on the degree to which the construction programme might be brought forward and undertook to consider their suggestions for accelerating preparation and procedures. But overall the comparison was not an unsatisfactory one given that the Department must follow proper procedures and has to temper established need with priorities and resources. The pace of construction can be constrained by the statutory procedures which must be followed and which often lead to public inquiry. On average

there is a 12-year gap between planning and completing construction. However, the increased competitiveness and adaptability of the construction industry has recently contributed to improvements in performance. In 1984/85 of 26 major construction schemes completed 16 finished early and a further 15 schemes under construction at the end of that year were running ahead of schedule.

Maintenance
21. Assessing the physical state of the existing road network and maintaining it are also of utmost importance. But the Department has to take account of public expenditure constraints and the balance between maintenance and new construction projects, which bring environmental as well as economic benefits.

22. <u>Assessing Need</u> In the past few years better techniques have been developed to help in monitoring the physical state of the existing road network, deciding how to keep roads up to the required condition and judging priorities for the allocation of resources:

(a) Computerised Highway Assessment of Ratings and Treatment (CHART) records data of visible highway defects for processing by computer to identify lengths requiring treatment and priorities for action;
(b) Maintenance Assessment Rating and Costing for Highways (MARCH) provides in addition a preliminary costing of recommended treatment, used by some local authorities only;
(c) Deflectograph Surveys measure the remaining structural life of roads and suggest the treatment necessary to extend it;
(d) The Sideways Force Coefficient Routine Investigation Machine assesses the need for surface dressing treatment;
(e) The National Road Maintenance Condition Survey (NRMCS) is an annual exercise of visual surveys of 9,000 sample sites sponsored by the Department and local authorities.

Improvements to these are under active consideration all the time.

23. <u>National Motorways</u> These are first priority for maintenance. Although they represent less than 1% of the road network they carry about 28% of road freight traffic. Some 5% of motorways can be expected to need renewal each year, that is to say 110 kilometres. In addition, there is a backlog of some 104 kilometres accumulated during the 1970s when priority was not given to them. In 1985/86 115 kilometres are due to be renewed at a cost of about £70m. From 1986/87 128 kilometres will again be programmed, subject to the availability of funds, so as to eliminate the backlog within the next 5 or 6 years.

24. <u>National Roads</u> On national roads the estimated need is

CENTRAL AND LOCAL GOVERNMENT POLICIES

for the equivalent of about 300 kilometres of single carriageway to be renewed each year to keep pace with deterioration. Some 160 kilometres will be renewed in 1985/86 at a cost of £31m. There is a backlog of about 480 kilometres, 4% of the total as at the period ending March 1986. The elimination of this will depend on the availability of additional funds, although it is recognised that the need will increase meanwhile and work cost more.

25. <u>Private Sector Involvement</u> The Department has for some time been looking at ways of increasing the involvement of the private sector in highways maintenance. At present in almost all cases maintenance of national roads is carried out by the appropriate county council or London borough as the Department's agent. After abolition this will change when some motorway maintenance will be contracted out to the private sector. But the appropriate district will generally become the Department's agent for national road maintenance. Local authority direct labour organisations have also come under increasing pressure to become more competitive. The 1980 Local Government, Planning and Land Act introduced a regime of competitive tendering and separate accounting for work carried out by local authorities. Regulations since then have gradually subjected more work to competitive tender.

26. <u>Minimising Disruption</u> In recent years the Department has made considerable strides in improving traffic flows and safety at major road repair sites. The contra-flow system has been progressively refined with better cross-over designs which allow smoother, safer and faster changes of lane. Delays at works sites have also been reduced by the introduction of lane-rental contracts. These contracts are making significant in-roads into contract times and are saving the nation considerable sums in delay-costs. The Department estimates that the savings to road users on the 4 experimental contracts in 1984/85 amounted to about £1m. The figure for the current year should be substantially higher than that. In the longer term recently announced improvements to the specification of new and reconstructed roads will mean longer intervals between major renewals. The Department has recently published a Code of Practice for Routine Maintenance of Motorways and National Roads and has introduced 5-year rolling programmes for major renewal schemes. There may be some scope for consultants to be involved in the implementation of the system particularly in the collection of highways inventory data.

LOCAL AUTHORITY ROADS

27. The 79 local highway authorities are responsible for 96% of the total road network in England. This includes 25,000 kilometres of principal roads, more than twice the length of national roads. The Government has no direct involvement in or control over their road programmes but is concerned that those of their roads which match national roads in traffic are of similar quality and standard.

Local Authority Priorities

28. Highways are part of a local authority's overall land-use planning strategy. The framework is provided by development plans. Structure Plans give a general indication of where development should be located and its scale. Local Plans provide detailed locations for the future supply of land for housing and industry, areas of restraint and development programmes. Both include proposals on the development of the road network and other transport services. The local authority also has to take into account the balance it wishes to strike between construction and maintenance, and the relative priorities of all the services it must provide within the resources at its disposal.

Government Objectives

29. Local authorities provide a wide range of services including roads. Their expenditure has considerable impact on the economy and must be covered by government spending plans. The Government is also concerned about the provision of a suitable strategic network for the country as a whole in which local authorities have roughly equal responsibilty. It is proper, therefore, that the Government should set out objectives for local roads. In providing resources the Government aims to encourage local authorities to provide and maintain roads on primary routes complementing national roads in quality, capacity and traffic; to improve urban roads to give industry and commerce better access, and to improve the quality of urban life by removing heavy traffic; to provide bypasses for other communities as a relief from heavy through traffic; and to continue to improve road safety.

Current Expenditure

30. Local authority current expenditure is supported from central funds through RSG block grant. Local authorities vary in their expenditure needs and the rateable resources from which to finance such needs. The objective of RSG is to supplement authorities' own rate income so that they can all provide the services for which they are responsible to similar standards, while charging a similar rate in the pound to their ratepayers. Block grant is distributed on the basis of grant related expenditure (GREs). The GRE for an authority is an assessment of the cost to that authority of providing a typical standard of service having regard to its general circumstances and responsibilities. It is not intended to be a normative level of expenditure: the amount actually spent on individual services is a matter for each local authority to decide. GREs (which are reviewed each year by the Secretary of State, in consultation with local authority Associations) are calculated by reference to formulae made up of physical indicators. However, the RSG system ensures that less grant is available on spending above the total of all GREs which is a way of influencing authorities' expenditure. And there are now powers to "rate-cap" some very heavily overspending

authorities which have their maximum level of expenditure set by Government.

Capital Expenditure

31. Capital expenditure is financed mainly by borrowing but also from contributions from the current account and capital grants. The assumed financing costs of this borrowing attract RSG block grant through the GRE financing capital expenditure. Capital spending is regulated through capital allocations. Separate allocations for each main service, including transport, give authority to spend and borrow within certain limits. Local authorities are, however, free to switch allocations between services and to draw additionally on a set portion of the receipts from the sale of capital assets.

Capital Investment

32. Within the total resources available local highway authorities are entirely responsible for the level of capital investment in their roads. Although spending on road building in recent years has been less than the provision in Government expenditure plans the trend is to reduce this "underspend".

Table 6. Resources Available for Local Road Capital Investment

£m	1981/82	1982/83	1983/84	1984/85	1985/86
Provision	355	418	478	491	502
Expenditure	326	407	457	460(est)	-
Variation	-29(8%)	-11(3%)	-21(4%)	-31(6%)	-

33. *Transport Supplementary Grant* Central Government direct support is through Transport Supplementary Grant (TSG). The focus of this was changed for 1985/86 to capital investment in highways and traffic management. This was partly to encourage expenditure closer to provision (see Table 6) but also demonstrates the high priority given by Government to the PRN, bypasses, major urban roads and major links to the strategic network (roads of more than local importance). TSG is a block grant paid at a flat rate, currently 50%, on estimates of expenditure submitted by local authorities. It is distributed in an annual settlement. The Secretary of State's decisions about the apportionment of the grant and about capital allocations are made on the basis of Transport Policies and Programmes (TPPs) called for each spring by a departmental circular and submitted each summer by the local authorities. TPPs explain an authority's policies for improving its roads and set out its proposed capital programme. In deciding how much of an authority's estimated expenditure to accept the Secretary of State considers the priority allocated to schemes by the authority, compares them with other authorities' programmes, the value for money they represent and the extent to which they relate to roads of more than local importance. Once the annual settlement has been approved by Parliament no changes can be made to the amounts of grant.

34. **1985/86 Settlement** In the 1985/86 settlement £160m of grant was given at a flat rate of 50% on £320m of expenditure out of a provision of £502m (Table 6). This enabled work to continue on 214 major (over £1m) schemes and to start on 75 major schemes. Local authorities are free to use the remaining £182m of capital allocation on any expenditure on their roads.

35. **Investment Need** For the moment the main indicator of the desired expenditure on local roads is the sum of bids submitted in TPPs. In 1985/86 these came to £740m. The apparently unmet bid of £238m may not all really represent unsatisfied demand. Most authorities, often for good reasons, bid for more than they can reasonably expect to get and in many cases to achieve. Their perception of need also varies according to the policies they are pursuing at any given time. In 1983/84 they bid for £702m, had £478m accepted for support and are estimated to have spent £457m on capital programmes (Table 6).

36. **Pace of Road Building** The CBI's report "Fabric of the Nation" found some £2.07bn (at November 1982 prices) of improvement on specified local authority roads to be needed. The Department's information was that £0.73bn of investment was planned at estimated outturn price. Local authorities, of course, decide on programmes and expenditure themselves. They need to judge capital road schemes against other priorities and must act within the total resources available and in accordance with the Government's control systems. They have to decide whether to maintain or build. Like the Department they must follow statutory procedures which can cause significant delay. Annual budgeting causes problems especially over continuity and assurance of government funding on lumpy major projects.

37. **Planning Ahead** It is true that the annual cycle of resources and grant decisions means that no assurances can be given of forward commitment by Government to any particular level of capital allocations or TSG. However, since the public expenditure resource planning system works on a 3-year basis, for planning purposes local authorities have some idea of the overall level of resources which will be available over that period. Also the Department commits itself to continuing to provide capital allocations and grant for the major schemes identified and named in programmes accepted for TSG. It is also taking other steps to increase confidence in forward planning. It will indicate without commitment whether a scheme being planned is likely to be eligible for TSG and will conditionally accept for TSG some larger schemes well before construction mainly where there are major advance land and other costs. For 1987/88 and onwards it will call for long term programmes to encourage forward planning and to give a clear context for schemes for which grant is sought. It is reviewing, jointly with the local authorities, the investment needs of their principal roads up to the end of the century including those on the PRN.

CENTRAL AND LOCAL GOVERNMENT POLICIES

38. <u>Value for Money</u> The benefit to traffic and to the environment from investment in local authority roads is of the same order as that for national roads. Indeed, a recent rough analysis of a selection of schemes showed total benefits on new local authority roads of something like £2 for every £1 spent compared with £1.74 on national roads. But there are difficulties in achieving a standard assessment of local authority schemes. Some use COBA wherever possible, others a mixture of this and their own assessment techniques, and still others rely entirely on their own local criteria. The Department is discussing with local authority Associations the development of a common list of aspects which all can agree should be included in every case and to which individual local authorities can add local dimensions for their own planning purposes.

<u>Maintenance</u>
39. Both the Government and local authorities recognise the vital importance of the maintenance of local roads. It is supported through block grant. It is only eligible for TSG where a project constitutes a substantial improvement to a road satisfying the criteria for the grant. Local authorities considered a proposal by the Department that structural maintenance should be classified as capital expenditure as on national motorways and roads, but they decided against it because they felt it would not improve their ability to manage their capital programme. Provision in the Government's expenditure plans increased from £898m to £972m between 1984/85 and 1985/86, which is 8% in cash terms and well above the rate of inflation. Even so local authorities have in total recently tended to spend above the provision levels.

Table 7. Resources for Local Authority Road Maintenance

£m	1983/84	1984/85	1985/86
Provision	831	902	972
Outturn	877	949(est)	980(budget)

40. <u>Local Road Condition</u> The trend in general conditions is provided by the National Road Maintenance Condition Survey and local highway authorities are asked to give their views on the maintenance of their roads in their TPPs even though expenditure on it is not generally eligible. Otherwise there is no system for recording centrally the condition of local roads.

41. <u>The 1984 Survey</u> The 1984 National Road Maintenance Condition Survey shows that there has been some deterioration since 1980 but that no class of road is in significantly worse condition than in 1977 when the surveys began. Many local authorities claim that the relative stability shown by the survey report is misleading because basic defects are being concealed by thin courses of surface dressing, which is all they can afford, and there are engineering problems which

cannot be reflected in the survey. From a complementary deflection survey in which about 9 counties participate the report does, however, suggest that about 24% of principal road lengths has a residual life of under 10 years. This implies a need for 2.4% to be strengthened each year. This is more than the proportion strengthened in 5 out of the last 6 years. This suggests that a backlog is building up.

42. <u>Code of Good Practice</u>   In 1983 the local authority Associations published their Code of Good Practice for Highway Maintenance. This aims at achieving better value for money through improved management of maintenance, particularly by the adoption of preventative and remedial treatments as opposed to more cosmetic methods.

CONCLUSION

43. This paper has sought to give an overall view of the extent, usage, financing, programming and planning of the road-network in England. In Scotland and Wales responsibility is similarly shared by the respective Secretaries of State and the local authorities, and objectives are broadly the same. But financing and planning differ.

(a) In Scotland resources are allocated to local authorities on the basis of TPPs by the Secretary of State but there are no special grants for roads. Authorities fund their expenditure by borrowing and the costs of supporting this counts as relevant expenditure for the allocation of resources.
(b) In Wales there is no formal system of TPPs and resources are allocated to local authorities on the basis of simple highway plans. They include a special grant at a fixed rate of 50% for capital expenditure of more than £5m on road schemes and public transport. Otherwise resources are distributed by formula.

44. The position in England set out in this paper gives the context for the PRN. The 1983 White Paper "Policy for Roads in England" signalled:

(a) a review of the network;
(b) a review of the investment needs in the longer term, which was later extended to cover local principal roads off the PRN as well as on it.

These are in progress at the time of this paper.

4. A review of the primary route network within the UK

B. OLDRIDGE, Director of Transportation, Cambridgeshire County Council

SYNOPSIS. My paper will outline the initiative taken in the Eastern Region of the United Kingdom of a comprehensive review of the total Primary Route Network. It will give reasons why a comprehensive review was required, explain the criteria for choice, and the criteria for intervention levels for investment. The paper will outline how this concept was adopted by the CSS in 1982 as national policy. It will explain how the DTp, in partnership with CSS, agreed to a national review. Progress of this will be outlined.

Methods of traffic growth monitoring will be explained and financial resources necessary to secure completion of the Network by 2001, with 75% completion by 1990. The resources required will be compared with the PESC forecasts. An annual monitoring mechanism will be suggested. Methods of evaluation of the benefits of individual schemes, using COBA and the wider benefits that investment in Primary Route Network can bring, will be debated.

BACKGROUND

1. The Primary Route Network was a concept originally formed in the 1950's.
2. It has served its purpose well in directing long distance traffic onto specific routes. No comprehensive review had taken place since that date.
3. It became clear to the County Surveyors Society in 1979/80 in its overall review of 'Transport in the 80's' that a fundamental review of the Primary Route Network was needed in the light of macro changes in road transport, and industrial commercial and residential land use in the United Kingdom over the last 30 years. Also, major population changes, New Town developments, the emergence of the motorway network (which radically altered the pattern of movement), and the changing emphasis of port traffic, have all contributed to the need for a radical review of the network. Many roads which had great significance in the 1950's have become totally superseded in the 1980's and vice versa. In the light of these changing planning issues,

the County Surveyors' Society decided to embark on a review.

4. Members of the Society in the Eastern Region started the Review first and I was privileged to be the Chairman who steered the review in that Region.

In essence, the concepts of a Primary Network Review are startlingly simple. The steps taken were as follows:

1) Review the road network in the light of eligibility criteria.
2) Assess growth factors in the Region.
3) Assess improvements needed in the light of traffic growth to 2001.
4) Calculate a programme of investment, based on overload factors.
5) Monitor growth and progress annually on a regional basis.

Naturally, within the overall framework individual schemes would be required to comply with COBA in order to enter a firm programme in either Trunk or County road budgets.

THE REVIEW IN THE EASTERN REGION

The review was a partnership between Counties and the Department of Transport and a Working Party, reporting jointly to the Director DTp. and County Surveyors, was established to carry out the review.

5. The Terms of Reference were as follows:-

"To consider:
(i) The definitive Primary Route Network for the Eastern Region.
(ii) Growth indicators in the Region
(iii) Longer term needs for improvement to the network.

In addition to defining what the long term Primary Route Network should be, it made recommendations on the investment necessary to make each route suitable for the needs of its traffic and the communities through which it passes. For this purpose, the Report took account of existing trunk and county programmes and also identified those sections of route where it seemed likely that environmental, financial or local political constraints make the achievement of a satisfactory standard particularly difficult.

6. In drawing up the Report, the Working Party considered the evidence of traffic growth, both past and future, in relation to national and regional trends in population, gross domestic product, growth of ports traffic and any development factors which are significant in land use planning terms. They also considered whether there was a need for further traffic monitoring over and above the information available from national and other traffic surveys.

7. The Department of Transport defines a primary route as one not being a route formed by any part of a motorway, which the Secretary of State is of the opinion after consideration with the Highway Authority concerned, will

provide the most satisfactory route for through traffic between two or more places of traffic importance. A more detailed interpretation is included in paragraph 4 of the Traffic Signs Regulations and General Direction 1981 (SI 1981 No. 859). In combination with the motorways, the primary routes provide a national network which serve the needs of through traffic travelling between different areas of the country. Motorways have their own distinctive blue-backed signs, and primary routes are made easily recognisable to drivers by using signs having a standard green background. All these signs have a limited and defined set of forward destinations for travel along the route.

8. The first task was to review the network itself and delete those roads which were unnecessary and insert roads which had grown in importance. It is clearly important to chose standard criteria. It was decided that generally all towns of more than 25,000 population should be linked to the Primary Route Network. This task was carried out with very few boundary difficulties and the network was accepted by DTp.

9. The second task was to measure carefully all those planning factors which influence traffic growth and traffic patterns in the region. This was particularly important because of the high level of economic activity and growth in the region. 'Growth indicators' were measured and percentages compared with national trends from 1976 - 1991 and 1976 - 2001 for the following factors:-

1. Population.
2. Households.
3. Employment (employed residents growth in jobs)
4. Car ownership - Cars per household
5. Trips and Ports related traffic.
6. Various screen line counts.

The data is based on Office of Population Census and Surveys (OPCS) 1977 based population projections, Dept. of Environment household projection and Department of Employment data.

TRAFFIC FORECASTING

10. Forecasts of traffic flows on each part of the network were needed to establish a basis for assessing what improvement was necessary and hence the appropriate investment needs for each route.

In some areas of the Region, traffic forecasting models, developed for specific projects such as for M25, were used to produce forecasts. In areas where specific models did not exist, forecasts were produced by obtaining the latest surveyed flows factored to 2001 (Low and High Growth), using factors for which the derivation is discussed later.

11. The surveyed flows were obtained from Department of Transport sources and from each county in the Region.

CENTRAL AND LOCAL GOVERNMENT POLICIES

The flows were converted to the common base of 1981, (Annual Average Weekday (AAWD) for 1981 (16 hr)).

More will be said later in paragraph 22 about a new and comprehensive annual monitoring of traffic growth on the whole network.

DERIVATION OF GROWTH FACTORS

12. The first step in producing the growth factors was to use the growth in vehicle-kilometres as given in the 1980 National Road Traffic Forecasts and break them down first to each Region and then to each County, using the trip end forecasts. This gave factors which showed how growth in traffic for each county compared with national figures.

13. The method of breaking down or disaggregation is given by the following calculation:-

$$\text{Local Vehicle km factor} = \frac{\text{NRTF 01}}{\text{NRTF 81}} \times \frac{(tc^{01}/tc^{81})}{(T\ 01/\ T\ 81)}$$

Where NRTF 01 is the vehicle km index for 2001

tc^{01} is the total car driver trip end for 2001 in County.

and T^{01} is the total car driver trip end for England in 2001.

The measures of national growth in vehicle kilometres are:-

	Low growth	Actual	High Growth
1978	–	0.97	–
1981	–	1.0	–
1991	1.126	–	1.320
2001	1.219	–	1.563

INVESTMENT

14. Following identification of traffic growth data we were then in a position to identify investment needs and a possible programme up to the turn of the century. Because of the buoyant nature of the Eastern Region economy, it was realised that growth was not a 'straight line' linear increase and that approximately 75% of the total investment to the year 2001 would be required by 1991.

15. The first stage in identifying the necessary investment on County Primary roads was to identify for each length of road the design flow threshold above which the road would need to be improved. Whilst current practice does not recognise a particular standard procedure for calculating the capacity of an existing road, the threshold was determined using the guidance given by the Department of Transport's Standard: Road Layout and Geometry: Highway Link Design (TD 9/81) which sets out the design flows for new roads.

16. Having identified which lengths of road would be in need of improvement on a common basis throughout the region, for both low and high traffic growth, the next stage was to determine the design standard appropriate to that improvement. The 2001 predictions and TD 9/81 recommendations were used as the basis for this.

17. The final stage of this work was to cost each improvement. Where there was already a preliminary or detailed design for a scheme (eg. schemes with start dates in early years), then the known estimated cost of the scheme was used. If no such costs were known, then the scheme cost was calculated using unit rates derived from similar schemes. Because of different land costs, terrain, soil conditions, etc., within the Region, each County determined its own unit rates and costings rather than applying a set of standard rates.

18. Trunk Road Assessment was arrived at in a slightly different way. Needs to the year 2001 was assessed after consultation with all Counties, and taking into consideration the Medium Term Plan Programme based on White Paper schemes and County schemes receiving 100% grant.

19. It is not the purpose of this paper to explain in detail the programme of work which was identified in the Eastern Region. Sufficient to say that it totalled £1,235M Because of the very high levels of growth evident in all of the planning indicators in the region, particularly in the early part of the plan period, it was agreed to target for completion of 75% of the programme by 1991.

MONITORING

20. As in all good management plans, it is vital that the plan and programme is regularly and formally monitored to review progress and make any necessary adjustments. Two separate monitoring processes have therefore been established. The first is to monitor annually the actual traffic growth on the regions total Primary Road Network and the second to monitor progress against targets.

21. A formal agreement between all Counties in the region and the DTp. to monitor the Primary Route Network by supplementing the national census points has been established.

22. The Department of Transport has, since 1979, been operating two compatible censuses in England, Scotland and Wales, to provide a national data base. Their Link Based Census provides comprehensive information on traffic flow levels as every major road link is counted once in a 6 year survey cycle.

Their Core Census is designed to give estimates of national traffic trends by type of vehicle and class of road. Sites have been randomly chosen throughout Great Britain in such a way that the distribution of sites, for each road class, between Scotland, Wales and the regions of England, is proportional to the length of road. This ensures that, whatever variation between regions there may be in traffic trends, each region is given an appropriate weight in the national estimates.

23. The sample size of the Department of Transport's Core Census is not large enough to give acceptable estimates at a regional level. The Eastern Region of the County Surveyors' Society has set up, in partnership with the Department of Transport, a system whereby the randomly selected Core Sites are supplemented by additional sites within each county in the Region, selected by the same method of randomisation. In return for providing this additional information, the Counties receive annual statistics giving traffic growth on the Region's Primary Road Network.

24. Secondly, the programme and investment levels measured in outturn cash flow are surveyed annually on both the Trunk Road and County Primary elements of the network and compared with targets. Both monitoring reports are brought together in a single report which is agreed between the Director DTp. and all County Surveyors. It is then presented to the Minster of State for Transport at the annual Eastern Region Consultative Committee. The original Eastern Region Primary Route Network Review document was produced in November, 1983 and two monitoring reports have now been prepared in 1984 and 1985.

25. The whole management process of Primary Route Network planning and monitoring corporately in the region, is therefore well and truly launched. However, because the concepts follow the classical theory of all management planning, from development of standard criteria, through plan formulation, to programme acceptance and monitoring, it all sounds very easy to implement.

26. Nothing could be further from the truth! The successful launch and control of the package has demanded a high degree of flexibility and goodwill, both professionally and politically; and a willingness to work corporately for the success of the region as a total identity. Particular mention must go to the Standing Joint Committee of County and DTp. officers who do the bulk of the work and research.

27. The main purpose of Primary Routes is to assist growth in the economy by lowering the unit costs of road transport for longer distance traffic. In order to do this properly, the network must ignore cross boundary problems and be assessed for investment on a wider canvas. Perhaps the best unit is regional, which in turn fits in with the concept that regions and not smaller geographical units, have economic identities. It follows, therefore, that Primary Route Network planning would be more accurately assessed using regional traffic models and growth factors. I would argue that perhaps the time has come for the

acceptance by DTp. that regional traffic growth is a more accurate barometer of need, than the national high and low forecasts.

THE NATIONAL REVIEW.

28. Having successfully operated the planning concepts at Regional level for 3 years, the question is: 'Can it be completed nationally'? I mentioned at the start of the paper that the County Surveyors' Society published a major document in 1982 entitled 'Transport in the 80's' which through a series of papers, sought to comprehensively review the whole field of transportation policy for the decade ahead. Paper No. 4 set out the methodology for the review (which was the methodology explained earlier in the paper, successfully tried as a pilot study in Eastern Region), and published a map showing the Societies suggestion for a revised Primary Route Network.

29. The Minister of State for Transport announced in August, 1985, that she intended to review the Primary Route Network in England and published, for consultation purposes, a map of a revised Primary Route Network. Unfortunately, there does not appear to be a common criteria for inclusion of routes in the network, which has led to some disparities in the density of the network across the country. This could prove to be a serious failing if it is not corrected. The revised T.S.G. eligibility criteria which gives preference to Primary Routes, has made it imperative that the network is standardised to achieve fair distribution of resources.

30. In spite of the problems, it must be said that we are now moving in the right direction - albeit a little belatedly - and the differences I am sure can be reconciled between the County Surveyors Society 1982 map and the 1985 DTp. consultation document, particularly if common criteria for inclusion of routes can be accepted.

31. The guiding principles in the DTp. review are:

(a) retain the Primary Route Network as the system of routes serving longer journeys - rather than the needs of local traffic - between places of traffic importance, primary destinations;

(b) adjust the Primary Route Network only where there are compelling reasons because changes are confusing to the motorist and expensive to implement.

It is hoped to finalise the network review by the end of the year. Important though the network review is, it is only a first step on the total management package. Indeed, a network review alone would merely be an exercise in re-signing. Investment in the network to bring it up to the correct design standards for expected traffic to the turn of the century, should be the main purpose of the review. It is hoped that the DTp. will extend the exercise, in partnership with County Councils,

CENTRAL AND LOCAL GOVERNMENT POLICIES

to assess the total needs and compare it with PESC levels of spend. The criteria for carrying out the full exercise were tried and tested in the Eastern Region (set out in paragraph 2) and have been accepted by the Minister of State as a corporate planning document for that Region. It is now essential that the review, on similar lines, is completed nationally.

32. In 1982, the County Surveyors Society made a first rather crude global attempt to assess the resources needed to improve the whole network. The shortfall when compared with total PESC allocation for Trunk and County Primary Routes was of the order of 15% in 1982 which was not an unsurmountable gap when the economic benefits were considered. The 15% shortfall in the financial year 1981/2 was £68M per annum or 2.5% of the total planned expenditure on Road Transport.

33. Finally, I cannot resist a few words about evaluation of benefits and the shortcomings in the COBA technique. Let me preface my remarks by saying that COBA is a unique tool for assessing priorities on a common basis and over the years has been invaluable for that purpose. No doubt Ministers of other departments must look with envy on the techniques used in DTp. because it does give a very fair comparison of schemes for investment priorities.

34. Whilst it is an excellent comparative tool, it is a poor absolute one and indeed, many transport planners (including the County Surveyors Society in the report 'Transport in the 80's') accept that it seriously underestimates the benefits to be gained from major investment in complete networks. The concept of 'additionality of benefits' comes into the equation as later schemes are brought into the network in the same way that the last half dozen pieces in a jigsaw puzzle often bring to life the whole picture.

35. I personally used this argument of 'additionality' when giving evidence at a recent Public Inquiry. My argument was that the scheme was the last link in a dual carriageway network of over 200 miles and as a result, the benefits were out of all proportion to the COBA assessment for the scheme when judged in isolation.

36. The macro benefits of schemes which are the final parts of a network are not properly assessed, although the CBI, and FTA have both acknowledged its existence. 'Additionality' is, of course, judgemental, and open to possible exaggeration. Nevertheless, it is real - and my hope is that this additional benefit will be accepted to justify the extra resources needed to complete the Primary Route Network over and above PESC levels.

Discussion on Papers 3–4

PAPER 3   OVERALL REVIEW OF NATIONAL NETWORK
TABLES UP-DATED TO MARCH 1986

Table 2.   All Motor Vehicle Traffic (million vehicle kms) 1983 (latest available)

National	Principal Primary	Principal Non-Primary	Total	% of Primary Traffic on National Roads	Primary Traffic as % of Principal Roads
39,600	41,500	47,000	128,000	49	47

NOTE: The first figure in the final sentence of paragraph 3 then changes from "40%" to "50%".

Table 4.   Transport Expenditure (mainly roads and bus/rail subsidies)

	Outturn £m			Planning £m	
	Final	Provisional	Estimate		
	1983/84	1984/85	1985/86	1986/87	1987/88
Central Government	1,713	1,864	2,151	2,430	2,440
Local Authorities	2,632	2,719	2,433	2,379	2,400
Total	4,345	4,583	4,584	4,809	4,840

Table 5. Roads Expenditure

	Outturn £m			Planning £m	
	1983/84	1984/85	1985/86	1986/87	1987/88
Central Government					
Capital	644[i]	721[i]	743[i]	816[i]	850[i]
Current	64	78	74	82	80
Total	708	799	817	898	930
Local Authorities					
Capital	457	447	494	533	1,670[ii]
Current	877	(est)946	977	1,115	
Total	1,334	1,393	1,471	1,648	1,670[ii]
Total All Roads	2,042	2,192	2,288	2,546	2,600[ii]

i. This figure includes expenditure on structural renewal, which is classified as capital. Road construction accounts for £514m in 1983/84, £568m in 1984/85 and £599m in 1985/86.
ii. Overall planning totals as at 1985 subject to revision. The figures for local authorities are notional and based on the assumption that roads spending accounts for the same proportion of provision as in 1985/86.

Table 6. Resources Available for Local Road Capital Investment

£m	1981/82	1982/83	1983/84	1984/85	1985/86	1986/87
Provision	355	418	478	491	502	533
Expenditure	326	409	457	447(est)	494(est)	−
Variation	−29(8%)	− 9(2%)	−21(4%)	−44(9%)	− 6(1%)	−

Table 7. Resources for Local Authority Road Maintenance

£m	1983/84	1984/85	1985/86	1986/87
Provision	844	902	972	1,115
Outturn	877	949(est)	977(budget)	N/A

NOTE: The increase in provision in cash terms 1985/86 to 1986/87 is thus 15%.

DISCUSSION

MR J. R. ELLIOTT, Greater London Council
With reference to Paper 3, there would seem to have been a reversal of roles between Mr Butler and myself: the last time I was in consultation with him was when we wanted a small road and the DOE was against its construction on the grounds that unacceptable environmental damage would be caused by the extra traffic. Now the DTp seems to be proposing a number of large new roads for London which, as a result of the traffic generated, would do serious environmental damage to the city as a whole.

The primary road network consists of many hundreds of miles of ordinary roads in London, such as the Fulham Road. The widening of such roads in order to make them similar in 'quality and standard' to the national roads, which they 'match' in traffic terms, is clearly impossible. Furthermore, such a policy could very well make travel worse within the capital. In terms of transport, London is definitely very different from anywhere else in the country.

Two further points in Mr Butler's Paper that need consideration are as follows.

Firstly, great weight is given to value for money in a COBA type analysis. The national programme shows a benefit to cost ratio of 1.74:1; some of the roads currently being proposed for the GLC area have ratios less then 1:1. Surely, if we are really concerned with value for money, we should be spending our money on matters that provide higher returns, that is, accident remedial schemes, public transport subsidies, or even non-corrosive salt (a reference to a recent letter in New Civil Engineer). This would be in keeping with Mr Oldridge's comment on the need for a total management package.

Secondly, it is interesting to note that, in paragraph 39 of the Paper, there seems to be some criticism directed at the local authorities for their having spent over the provision levels on maintenance, while, in paragraph 41, it is recognised that a backlog is building up. Surely, we should be spending more on maintaining what infrastructure we already have rather than on building more roads requiring yet more maintenance.

MR M. N. T. COTTELL, County Surveyor, Kent County Council
With reference to Papers 3 and 4, few can doubt the wisdom of a primary route network. Indeed, the principle of trying to concentrate the longer distance trips on a few defined routes cannot be disputed. What is in dispute, however, is the best way to achieve that objective. As Mr Oldridge says, the existing primary route network was established at the start of the motorway era. Many designated primary routes no longer serve that purpose, while other routes, not green-backed sign routes, now perform that task.

The time has come to redefine the roads which showed the primary status, not only because of the advent of the motorway system but also because of the changing patterns of industrial traffic. However, to change the network will not, of itself, encourage vehicles to use it. We have to bring that network up to efficient standards. This must infer investment in the primary route network by both central and local government which, I believe, is essential to the productive base of our nation. At every meeting I have with industrialists, the need for a good efficient main road system is stressed, a view borne out by these two Papers. I echo the remarks already made that total public section investment is limited, but we must restate the case for roads for industry. Contrast the difference between the approach in France to links to the Channel Tunnel and our own meagre approach.

The impression gained from Mr Butler's Paper is that things are well under control and that problems are being sorted out. There seems to be no acknowledgement of the fact that a huge amount of work still needs to be done. In paragraph 35, the implication is that the capital allocation (nationally) for local roads is quite adequate, as local authorities always bid more than they expect, and do not spend all the allocation anyway. No account is taken of the possibility that bids might represent needs or that the financial control system in effect militates against overspending. The government constraint on capital allocation and the effect of changes on the annual revenue budget are not understood. As an example, Kent has been told that a TPP bid of £28 million is over-optimistic; an accepted capital expenditure of £15 million needs to be judged against an assessed need of £500 million.

The point also needs to be made, I think, that a single year acceptance is not sufficient for planning a highway improvement programme, and that what is needed is a commitment to increasing the programmes of work. As local authority road improvements give a better return on investment than national roads improvements (Paper 3, paragraph 38) this could be accomplished by a transfer of funds from national to local roads.

I share Mr Oldridge's concern (Paper 4, Paragraph 29) about the variability of the network. I think the differences in density may be a consequence of differing terms of reference in different regions. Certainly, in the South-East, we did not see the 'guiding principles' as quoted in paragraph 31, and these did not form part of the terms of reference. Rather than concentrate on minimal changes, we took a fresh look at what was needed. The density of our network is much less than in other parts of the country.

Further, if the sentence, 'It is hoped to finalise the network review by the end of the year', means by the end of 1986, then I would ask - Why the delay? The South-East report was available in May 1985, and, as far as I know, other regions were working to the same timetable. If it is to be another year before the network is finalised, how long will it be before the part of the study dealing with investment needs is published?

Much remains to be done. I believe that a sensible primary

route network, with a proper programme of investment, would substantially benefit the productivity of our industrialists. It is time that we grasped the nettle and helped to make UK Limited a more prosperous concern.

MR M. SPRINGETT, Travers Morgan & Partners
With respect to Papers 3 and 4, there are two points I would like to discuss. Firstly, Mr Butler's reference to 'value for money' as a criterion in determining priority for the development of the primary route network. Secondly, Mr Oldridge's reference to the 'additionality' of traffic benefits.

The example I take to demonstrate the points is in North Wales, namely the development of the A55 Trunk Road and the A470 Llandudno Connecting Road - a principal road within Gwynedd.

The A55 over a length of about 30 km from Llanddulas to Llanfairfechan is being built to dual two-lane cross-section, grade separated and also to standards set out in 'Roads in Rural Areas' and 'Roads in Urban Areas', as appropriate.

The Llandudno Connecting Road extends over a length of 5 km, connecting the new A55 at Glan Conwy with Llandudno. It replaces the A546 and is of 7.3-10 m cross-section and designed to Link Road standards TD9/81. Junctions are at grade. This has produced a low cost solution with horizontal and vertical alignments enabling minimum earthworks in spite of a very undulating terrain.

One major bridge is involved, which carries the new road over the Chester- Holyhead railway, and average construction costs are £1.26 m per km. The new Llandudno Connecting Road costs are 2.5 per cent of the total scheme costs. In terms of benefits, taking the road network as a whole, the A460 contributes 10 per cent. Obviously, the benefits derived from the A470 would not be accrued without the A55. Nevertheless, it does demonstrate the process of 'additionality' to which Mr Oldridge refers in his Paper.

My argument, therefore, is that the second tier of roads in the PRN is likely to provide good value for money for the following reasons:

(a) these roads will generally be less costly then first tier roads because the standard of facilities will be lower and more rigorous design standards will be applied;

(b) significant benefits will accrue not only in the usage of the second tier system, but also by providing better access to the first tier, on the one hand, and providing relief to it in some circumstances, on the other.

Therefore, I would conclude that for the immediate future, attention should be directed particularly towards the improvement of principal roads, most of which come within the control of the County Authorities.

MR M. P. HEYES, Greater Manchester Council
With reference to Paper 4, I would like to follow up Mr Oldridge's comments in respect of one of my favourite hobby horses - the great god 'COBA'.

The facts of the matter are that any economic analysis of a highway scheme, let alone an analysis of a total urban transport system, must be a very inexact and crude representation of an inherently complex situation. The further superimposition of the constraint of a standard national computerised methodology would result in an even narrower analysis of the benefits in any given situation. Also, and even more significantly, to allow administrators in the Treasury to use the results of this computation to determine completely the fortunes of individual schemes is, I believe, the single most inhibiting factor in the sensible development of the primary route network.

Economic analysis is an important tool, and the COBA methodology can be a useful starting point. However, it must not be a strait-jacket and, most importantly, it must not be allowed to maintain its current importance in the order of things, where it is exerting an all-powerful influence. If the Department of Transport is to get to grips with the major problems affecting London and the other large conurbations, it must profit from the lessons learned by the Metropolitan Counties, who have gone some way towards the implementation of comprehensive transport solutions with minimal local opposition. They have achieved this by being willing to appraise fairly all the likely benefits and any detrimental effects of a proposed scheme, and thus to arrive at a preferred solution.

I hope that the theme developed by the County Surveyors' Society, the Institution of Civil Engineers, and many others, will be taken on board by SACTRA when it makes its report. However, the single most effective move which could be instigated to speed up the implementation of the primary route network is to delegate more of the decision-making to the Department of Transport's regional offices. The latter are in a position to understand the local circumstances and they are at the sharp end of the investment programme, where it counts - that is, on the ground.

MR I. G. LAWSON, Strathclyde Regional Council
On behalf of Scots engineers who continue to practise in Scotland, I would first draw attention to the title of the Paper by Mr Oldridge. Although it refers to the United Kingdom, it applies, in practice, only to England and not to either Wales or Scotland.

Secondly, when we Scots get our act together, the results are magnificent - witness the events of two weeks ago at

Murrayfield. Unfortunately, although we have tried to
redefine our primary route network in Scotland along the lines
suggested by Mr Oldridge, the members of the County Surveyors'
Society in Scotland have not reached agreement on a common set
of criteria. For example, north of a line between Glasgow and
Dundee, there are only two towns with a population greater
than 25 000. At present, Campbeltown, with a population of
6000, lies at the end of a long cul-de-sac primary route. A
population-related criterion for defining a new primary
network might mean large parts of rural Scotland being 100-150
miles from the nearest primary route/trunk road. Meanwhile,
we have, at present, several different systems of classifying
roads - for signposting, for maintenance, for development
control purposes (as in Roads in Rural Areas). Could we
reconcile or rationalise these?

Finally, several speakers have referred to the fact that
funding improvements to the primary route network is currently
split between central and local government. Should the
development of the primary network not be the responsibility
of one authority - whether it be national or local government?

MR A. D. W. SMITH, Sir Owen Williams & Partners
In his Paper, Mr Oldridge has referred to the County Surveyors'
Society's paper, 'Transport in the Eighties' (published in
1981), with which I was associated. This review gave an
approximate estimate for the completion of a primary route
network by 1992 and, as Mr Oldridge has said in paragraph 32
of his Paper, the shortfall then was about 15 per cent of the
total PESC allocation for 1982.

The financial summary Table which appeared in the Society's
review is reproduced here, as I thought it would be
interesting to see how far its forecast is still valid today.
The Table's figures are in 1979 prices and it is difficult to
equate them to out-turn figures appearing in the Government's
White Paper. However, it would appear that, for trunk roads
and motorways, there was a shortfall of about £500 million, in
current prices, for new works for the period 1980-84. For the
period 1984-1992, the current allocation, if continued
annually, would just about keep pace with the figures given in
the Table, as long as not too much was switched to capital
renewal.

In the case of County roads, such an assessment is even more
difficult to make as not all the capital allocation is given
to primary routes. The County Surveyors' Society estimated 50
per cent. On this basis, the period 1980-1984 was at least
£200 million short. For the period 1984-1992, the total
requirement falls short by a figure which could be as large as
£2-3000 million, thus reiterating the original note, to which
reference is made at the foot or the Table.

I must stress that all these figures are very rough
estimates - I am sure that someone better qualified then I am
would be able to calculate them more accurately - but it is

quite clear that the end of the century is a more realistic target now than 1992.

Table 1. Cost of providing network to cater for 1992 traffic volumes (low growth). November 1979 prices £m. Percentage of total shown in brackets.

Shire Counties	1980-84 Programmed	1984-92	Total 1980-92
Motorway and Primary Trunk	1584* (79)	2196 (67)	3780 (72)
Primary County	429 (21)	1069 (33)	1498 (28)
Total	2013 (100)	3265 (100)	5278 (100)

Metropolitan Counties	1980-84 Programmed	1984-92	Total 1980-92
Motorway and Primary Trunk	142* (54)	320 (46)	462 (48)
Primary County	120 (46)	383 (54)	503 (52)
Total	262 (100)	703 (100)	965 (100)

Greater London Council	1980-84 Programmed	1984-92	Total 1980-92
Motorway and Primary Trunk	95* (58)	521 (21)	616 (23)
Primary County	68 (42)	1979 (79)	2047 (77)
Total	163 (100)	2500 (100)	2663 (100)

*Based on Programme A in "Policy for Roads: England 1980".

MR M. H. STRUTHERS, Department of Transport - Eastern Region
With reference to Paper 4, I would like to make the following comment.

The Department had not, in the past, shown very much interest in the primary route network, other than keeping an eye on the signing system. Following the initiative of the County Surveyors in the early 1980s, however, the Author's regional office had grasped the opportunity, with ministerial blessing, to assess the system and its needs within the eastern counties, and he had welcomed the chance to lead the joint Working Party which had been set up for the task. In carrying out this work, basic planning data had been used to establish traffic growth for each county and for each part of the network; and the resultant set of growth factors in the Report was sufficiently robust to be acceptable to both Department and counties in scheme assessment and cost benefit analysis. These growth factors were also of considerable value to other organisations, and the monitoring system which was established, and which Mr Oldridge continues to lead, will ensure that these are kept up to date. It is intended to repeat the assessments at roughly five-year intervals, as fresh census information becomes available, but the other regional traffic monitoring that is carried out will be an essential means of testing the validity of the forecast in years to come. On investment, indications so far are that for trunk roads it is close to the target, but that counties are falling behind, possibly as a result of the general constraint on available resources.

One point in relation to paragraph 16 of Mr Oldridge's Paper concerns the use of TD9/81 for the assessment of standards for improvement. This is essentially a means of looking at standards for the provision of a road to a new alignment. The selection of the actual standard to be used was left to each individual county, and was to be determined on the basis of the county's own practices and policies, and of local acceptability of particular traffic loadings.

MR BUTLER

A general view at this Conference, at least up to the discussion on Paper 3, seems to be that the primary route network (PRN) refers to an undefined main road network. This is, of course, not true. The network is clearly defined and distinctively marked by green-backed signs. It is determined by the Secretary of State for Transport after consultation with local highway authorities. It is not population related but designed to show the most suitable route for long distance travellers between places of traffic importance. Some such places have negligible population, for example, Scotch Corner on the A1.

The PRN is a joint responsibility, with the Department bearing all the cost of the trunk road element and the local highway authorities being responsible for the remainder with help from central funds. This is an effective partnership between local authorities, as highway authorities and agents, and the Department. There are no plans to change it. Rationalisation has been considered but the disadvantages outweighed the advantages.

PRN review

The 1983 White Paper on Roads signalled the intention of bringing the PRN up to date. It naturally follows that when the results of the review are announced, any re-signing which is considered appropriate to redirect traffic will be the responsibility of the relevant highway authority whose future investment plans should take this into account. It should be noted that the South East Area Review, involving all counties in the south east, set the guiding principle of changing the PRN only where there were compelling reasons to do so.

Some concern has been expressed that the Review is taking far too long. Consultation with local authority associations, following the completion of the joint regional reviews and approval in principle by Ministers, has been going disappointingly slowly and is still incomplete at the time of the Conference. It was made clear that the Department still awaited a response from one of the local authority associations, which had been due by the end of the previous October. This was eventually received in April. It is hoped to announce the revised network fairly soon now, so that implementation can begin.

DISCUSSION

The PRN reveiw specifically excluded the area within the M25, that is to say London's roads. There are no new road proposals in London but the Department has commissioned consultants to consider whether problems exist and, if so, what possible solutions could be employed. There is no presumption of a consistent quality and standard across all parts of the PRN. Standards of construction are determined according to the needs of the traffic using the roads.

Investment review

An initiative by the Eastern Regional Office of the Department of Transport and like minds in local authorities, in reviewing future PRN investment needs in their respective areas, was taken up nationally in 1984 and extended to include local authority principal roads off the PRN. Guidelines were set nationally and the same joint working groups as had worked on the PRN review conducted regional studies against them. This necessitated the assumption of a future PRN as a framework within which to work. It was not intended to represent a commitment either to the shape of the network or to levels of investment/resources. The future networks were the best guess of the engineers on the spot of how the PRN might best respond to the needs of future traffic. The regional reports are being collated into an appraisal covering the whole country, on which it is hoped shortly to consult local authority associations.

There appears to be a persistent belief that TSG support is directed exclusively towards the PRN. This is not the case in England where TSG is also paid on roads off the PRN. The criteria in Circular 2/86 demonstrate this, as does the make-up of TSG settlements. It is up to councils to provide adequate information to enable schemes to be properly assessed and compared, in terms of their benefits, against the criteria and according to their respective merits. We are currently seeking through a joint working group with the local authority associations to establish a common basis of assessment of individual schemes, and to improve the information we receive from authorities on their schemes, in order, in particular, to demonstrate the value of local road investment to help make the case for resources.

It has been claimed at the Conference that maintenance is ignored in favour of new construction. This is quite incorrect and must have resulted from a misinterpretation of Paper 3. In fact, provision for maintenance has been substantially increased (by 15%) for 1986/87, and certainly more is spent nationally on maintenance than on new construction.

Central government funding

Views have been expressed about the inadequacy of the national capital allocations. It has to be appreciated that resources are necessarily dependent upon what the country can afford. There can certainly be no question of achieving a greater proportion of the national resources for local roads unless councils can provide more adequate necessary detailed evidence of the benefits of local road investment. Only in this way can the case for resources for roads be measured against competition from other services.

The Government cannot commit itself to a predetermined level of resources years ahead, but the system of annual PES review on a rolling three-year cycle is intended to give confidence. Moreover, road schemes accepted for TSG are supported throughout the life of the schemes provided only that the local authority makes satisfactory progress on them. The long-term plans of councils are useful too and a necessary input to the consideration of the relative merits of programmes and to the negotiation in the PES round. Much depends also on councils' performance on schemes. It is worth remembering that councils who fail to deliver on their supported schemes are, in effect, not only depriving other councils of resources they could well use, but also making it difficult to justify the overall levels of resource cover requested.

COBA

On COBA, maximum value for money is being taken by some to be the only criterion. This is not so, although value for money is important. Some schemes are progressed for environmental reasons in spite of negative cost-benefits. In London, although high costs can reduce the economic value of some schemes, this does not mean that otherwise worthwhile schemes should not go ahead.

In spite of their complexity, urban schemes are capable of being assessed for their benefits by using exactly the same principles as for rural schemes. However, in urban areas the impacts are likely to be more complex and there are more elements in the framework. Nevertheless, fundamental COBA type considerations of time savings, operating costs, accident savings, to be measured against the investment needed to achieve them, are the same. A balance has to be struck between these economic impacts and the environmental effects within the framework. More work on the appraisal of urban schemes can follow the publication of SACTRA's report in due course.

5. The development of France's primary route network over 25 years

A. THIEBAULT, formerly Ingénieur Général des Ponts et Chaussées, Paris

SYNOPSIS. After a brief review of a few basic data on France, her transport needs, her road networks and her road traffic, attention is given to the very notion of "primary route network", which is to be understood as the system of main routes that carry the bulk of a nation's traffic. It so happens that in the case of France, this can be assimilated to the national road system, motorways of course included. The main aspects of the machinery for financing works on national roads is then examined, and the significance of Government agencies such as "the Plan" and "Datar" is explained. These agencies play an overriding role in France as to planning in the economical field on a national scale, and consequently in the field of road policy-making and national road programming. The toll policy on rural motorways is also explained. Successive comprehensive master schemes for the national road and motorway networks based on the two examples of the first master scheme dating back to 1960 and of the latest master scheme of 1985 are outlined. The development of the guiding principles behind these schemes is examined as well as the main issues concerning roads in France today.

I SOME BASIC DATA ON FRANCE AND HER TRANSPORT SYSTEMS

- 54 million inhabitants per 550,000 km2

- nowadays, 75 per cent of the population in towns (i.e. of more than 2,000 inhabitants) against 56 per cent in 1954.

- as of 1st Jan. 1985: 24 110, 000 motor vehicles
  of which 20 800, 000 are passenger cars
  and 3 310, 000 are heavy vehicles
  (lorries, buses and coaches).

The road network and the estimated traffic (1985)
(Lengths: in kilometers)
(A.D.T.: Average daily traffic: in vehicles per day)
(Traffic: movements in thousand million veh x km).

		Lengths	A.D.T.	Traffic
(i)	Motorways			
	rural area	4,843	16,800	28.2
	urban area	1,447	51,000	25.2
	total motorways	6,290	24,700	53.4
(ii)	National roads	28,228	7,400	62.4
(iii)	Total primary network	34,518	10,800	115.8

Moreover:

(iv) Local roads about of all kinds	1,150,000		about 180
(v) Grand total rounded up to	1,185,000		295

Just one word in passing: France has always been known and renowned for her extensive networks of local and secondary roads; but they have not been described in detail above as it is not to the point.

a - Development of traffic on the primary network.

From 1974 to 1982, overall traffic increased at an average rate of 2.4 per cent per annum, i.e. at half the annual rate of the heady days of the '70ies, prior to the energy crisis.

Up to the present, (September 1985), the year 1985 seems likely to keep in line with developments for 1983 and 1984, i.e. an overall increase of about 1 per cent per annum, zero-growth (if not a slight decrease) for the national roads, and an increase for the motorways (3 per cent per annum on toll motorways and 4.4 per cent on toll free motorways).

b - Passenger transport.

Nowadays in France, the road accounts for some 87.5 per cent of all modes of passenger transport, whilst rail takes up 11.3 per cent, and air traffic the remaining 1.2 per cent. Rail's share has constantly been decreasing over the years. At the beginning of the '70ies, road and rail had reached the same level. SNCF (French railways) now accounts for 88.5 per cent of all rail traffic: SNCF, Metro (Underground) and RER (express suburban lines) etc., and have thus undoubtedly a better share than railways in neighbouring countries. SNCF are nonetheless heavily in the red, and as the outlook for an improvement in goods transport is rather bleak, a huge effort is being made to slash expenditure and attract new passengers. Significant results have been achieved in this respect with the new high-speed TGV-line on the 400km stretch between Paris and Lyon: 5.3 billion passenger kilometers in 1983, and 7.7 in 1984; and this progression continues. More than half the trips are rated as induced traffic, which has exceeded all expectations.

c - Goods transport.
(billions tonnes kilometers)

	1964	1974	1984
Road	41.7 (30.8)	90 (45.3)	105 (52.5)
Rail	65.3 (48.3)	77 (38.7)	60 (30.0)
Pipelines	15.8 (11.7)	19.6 (9.8)	26 (13.0)
Inland water ways	12.5 (9.2)	12.1 (6.2)	9 (4.5)
Total	135.3 (100%)	198.7 (100%)	200 (100%)

Thus over 20 years freight movement rose from 135.3 to 200 billion tonnes kilometers and the total has remained roughly at the same overall level for ten years, thus putting a curb on former expansion. The road's share again increased from 30.8 to 52.5 per cent, but at the cost of other modes of transport. Transport by rail passed to a maximum in 1974, and since then has been declining, as a direct consequence of the crisis affecting the heavy industries.

Freight transport by road of course has an impact on road pavements, and on road programmes. Such has been the case for the French national road system since the great damage caused by the memorable winter of 1962/63 when large stretches of national roads had to be closed to heavy traffic for some 50 days. Since then, systematic annual programmes have been devoted to strengthening and modernizing existing road sections. Nowadays, more than two thirds of the national road system have thus been brought up to standard, and the remaining third is being undertaken, unfortunately at a reduced rate, as funds are decreasing.

d - Just a few words about some more questions.

About the tolls, a first question which is often raised abroad turns on the levels of the rates that are charged. As a matter of fact, there are for a given vehicle as many tariffs as motorway authorities, i.e. 10, and rates can now range from 1 to 2 (1 to 3 prior to 1983). For a passenger car, the average toll is 0.28 franc per kilometer; in other words, a rise in cost per km of about 70 per cent (of the cost of the fuel). For a lorry, the rates range from 0.43 to 0.60 franc per kilometer.

Traffic evasion is by no means as high as might be imagined; but it may take a certain time, possibly several months, until motorists and above all haulage contractors take the motorway in preference to the ordinary (and toll-free) roads.

One more figure: by the end of 1984, and on the whole of the 4.455 kms of toll motorways in operation at that time, on average, heavy traffic accounted for 17.4 per cent of total traffic.

The death toll from road traffic

At present in France 12,000 people are killed and 300,000 injured on roads per annum. This is for instance twice the U.K.'s rates of traffic accidents. Such figures, of course, cannot fail to have a bearing on French policies and programmes; and in fact they form the background to all road schemes. Though not dealt with in great detail in this paper, nevertheless the security aspect should never be overlooked when planning roads.

II - THE PURPOSES OF A PRIMARY ROUTE NETWORK SCHEME

A good question to start with, turns on what is meant by a "primary network" and what is the answer in the case of France. As a matter of fact, such a question was raised at the beginning of the '70ies. At that time, the national road system was drastically cut from some 80,000 kms to a small 30,000 km network, the rest being transferred to 95 grant-aided local governments. The issue at the time was how to determine which among the 80,000 kms were the sole routes of real national significance: a rather controversial matter as one may easily imagine, and a rather long story. To cut it short, it was eventually recognized that the primary route network, where the state was directly concerned, in other words entailing an unquestionable significance at national level, could be nothing other than (i) all road interconnections between towns and built-up areas of more than 50,000 inhabitants along with (ii) main routes representing a special interest at national level in terms of development plans. This has lead to a new classification of roads, including a network of some 28,000 kms of national roads on which everybody eventually agreed and which is still in force today in my country along with some 6,520 kms of motorways, as will be seen further on in this paper.

As to a master scheme as an overall framework for this national road system, its usefulness is so obvious that there is no point here in going into further detail. For there are, broadly speaking at least three main reasons for drawing up such a scheme: (i) at state level, it is likely to ease annual discussions with the Department of Finance about the budget for national roads on the basis of a comprehensive master plan presupposing continuity over the years; (ii) the scheme also permits the attainment of better coherence within the objectives of the various sectors of activity as to the use of public funds; and (iii) inside a given region, where decision-making implies taking into account the various purposes of the state, the master scheme provides local governments and authorities with the required data on the national road network. Thus it allows a correct overall adjustment of their development plans, whilst providing the private initiative with information on Government's views as to motorway and national road building (which may prove of great help for instance in the case of relocation of activities). These three objectives are extremely well known nowadays, so much so that it would be be felt as a case of misgovernment and maladministration should the state fail to draw up and release such a master scheme. But again, here it should not be forgotten that before 1960 we had no such master plan for our national road system. Of course, the problems were not the same as those in previous years, since widescale urbanization and industrialization had not yet arrived.

III - FINANCING THE NATIONAL ROAD NETWORK

There are broadly speaking four main sources of funds for the national network. i.e. the national roads and the motorways:
 (a) the state's annual budget and its "special funds",
 (b) "contributed funds",
 (c) funds raised by public loans,
 (d) tolls.

To give some impression of an overall idea of the questions, the total budget for the national road network amounts to about 18.4 billion francs (i.e. about £1,533 million) for 1985 - and the shares of the 4 sources are respectively 46.4 - 16.8 - 26.7 - 10.1 per cent of the total.

For the sake of clarity, it is convenient to join (a) and (b) above, which both refer to the state budget discussed in paragraph A which follows; and in the same way, join (c) and (d), which both turn on the issue of the toll policy dealt with in paragraph B.

A   THE STATE BUDGET

Until the early '50ies, the national road system in France had always been financed by the annual budget of the state. In 1951, it became clear that the scale of needs together with the urgency of meeting the requirements of ever-growing traffic demanded new sources of finance.

Among those new sources, mention should be made here of "contributed funds" made up of money contributed by local and regional governments towards costs associated with work on the national road system. We will have to go back to that method of financing later on, as it has become a major issue in recent years, since the inception of the policy of "plan contracts".

Even more significant has been the institution at the end of 1951 of the "special road investment fund", set up on the basis of an additional specific tax of a few centimes per litre of motor fuel. A certain percentage (at the outset, 20 per cent) of that tax was to be earmarked for road expenditure. In the beginning that special fund was to be exempt from the usually painstaking discussions about the state budget; but eventually, and under the pressure of urgent needs elsewhere, that proved to be unfeasible. Moreover, in the eyes of the Department of Finance it would have been seen as an unjustified infringement of the sacrosanct principle of the unicity of the state budget. Despite a cut by half in the percentage allotted to roads, this fund has proved to be of great help. Since its inception and up to the early '80ies, it has borne the bulk of road expenditure on the national network.

In 1982, that source of financing was complemented by a very similar recipe with the "special fund for big works" which was set up to meet the new requirements of the grand projects envisaged by the President of the Republic during his seven-year term. Basically, the same main idea: again a specific and additional tax on motor fuel (a tax of one centime more per litre nowadays yields some 380 million francs!); and again a special fund that has been anything but specific. Even so, some of the money has been allocated to roads; but the lion's share has been allotted to rail, among other projects, to the new high speed train, TGV-Atlantique, (from Paris to Le Mans and Tours) and to various urban transit systems such as new underground or tramway lines; a clear sign that roads can no longer claim to be the main issue in the field of transport.

B - THE TOLL POLICY

At the end of the '50ies, there were virtually no motorways in France (130 km, as of Jan.1960); despite an increase each year in the special road investment fund, the annual rate of motorway construction has remained below an average of 100 kms, so at that rate it would have required some 20 years to achieve the first 1,828 km motorway programme that had just been approved by the cabinet of that time. The need and the urgency to obtain finance from sources other than those of the state budget became one of the mainstreams of political concern. Borrowing money by means of public loans appeared to be the sole way out of the dilemma; that meant imposing tolls on motorway users. A Hobson's choice indeed: no tolls but motorway construction at a snail's pace, or tolls, as on the first test toll motorway which had precisely just been opened to traffic on the Riviera in 1961, and then an outlook of possibly several hundred kilometers each year (1981: 419 kms were actually completed).

After a controversy over where tolls should be levied (everywhere or only in urban areas, or else in rural areas), collecting tolls on urban motorways appeared, on second thought, as "politically too unsaleable", and the solution that eventually prevailed favoured tolls on rural motorways only.

Two stages can be distinguished here: (i) in the early '60ies, 6 public motorway authorities (in fact subsidiaries of The National Savings Bank) were set up, each receiving a 30-year concession for the construction and operation of some 500 kms of motorways; (ii) and in the '70ies, at a time when that system of public concessions appeared to be tailing off financially speaking (those 6 public authorities were rather reluctant to take on new motorway stretches, which inevitably were to carry less traffic), and there was a drift towards privatization. Following a memorable competition between public and private sectors, 4 additional private motorway companies (contractors backed by bankers) were set up. And that system of 10 motorway authorities has undoubtedly led to considerable achievements, with nowadays some 4,500 kms of toll motorways in operation. But that system in its turn reached its limits about the beginning of the '80ies: the increase in construction costs (which have quadrupled over the last ten years), along with a worsening of loan conditions (with interest rates multiplied by 3 since the '60ies) meant that 3 out of the 4 private motorway companies were led rapidly to the brink of bankruptcy: not enough traffic, and therefore not enough money from the tolls, and finally not enough cash to meet their financial commitments.

After the landslide election of 1981, when France turned left, the issue at first was how to do away with those tolls; on second thoughts, and as buying out all the motorway authorities appeared as financially unaffordable, two main objectives were set up in 1982: (i) a gradual merging of the private motorway companies with the existing public motorway authorities and (ii) a gradual doing away with discrepancies at the tolls (in 1981: from 1 to 3 for a given vehicle, depending on the motorway), possibly together with a curbing of toll rates.

As to the question which may occur to the foreign reader on how those private companies have constructed, maintained and operated their motorways, to tell the truth, I must confess, one can hardly distinguish between them and the motorways of the public sector. On reflection, I can hardly believe this is due to the specifications that were set up in the concession covenants. It's rather that all those private companies made a point of doing things in quite the same way as the public motorway authorities.

IV - THE MACHINERY

It goes without saying that the Department of Transport, in particular with its directorate of roads and its central technical sections, plays an outstanding role in the drafting of road programmes. Less known abroad may be the roles therein of the "Plan" and of "Datar".

"The Plan" can hardly be disconnected from any public planning consideration in France, nor any field of activity, roads included. The "Plan" is the Government planning organizations set up by General de Gaulle in 1947 to direct postwar reconstruction and put together France's First Plan for 1947/52 by way of a cooperative effort involving long meetings with civil servants, businessmen, union leaders, etc. In short, the impact and success within government circles were such that the "Plan" has outlived all the political changes and has become a fully-fledged ministry which now leads all the machinery of state policies in the economical field (including, among others, road issues). But unlike planning organizations as they are known in eastern countries, the "Plan" in France is anything but constraining: it builds upon mutual agreement and concerted effort; and experience shows that it works wonderfully well. We are now at the end of the 8th Plan (1981-85) and we are preparing for the 9th; and all those plans include the road schemes we are going to speak about in the paragraphs that follow.

"Datar" is an acronym for the French words "délégation à l'aménagement du territoire et à l'action régionale". In plain English, this could be rendered by (governmental) delegation for development planning and regional action: that name clearly shows what it is about and it is pointless to elaborate. Making forecasts of future requirements for developing areas of course implies drafting, among others, the road schemes to that effect: so it is quite understandable that "Datar" have their say in road schemes drawn up by the Ministry of Transport.

V - A BRIEF OUTLINE OF THE DEVELOPMENT OF THE PRIMARY ROUTE NETWORK IN FRANCE SINCE WORLD WAR II

Looking back, 4 main periods clearly stand out:

(i) up to the early '50ies:
the age of postwar reconstruction along with a rapid increase in motorization, as in all western countries.
(ii) in the late '50ies and at the beginning of the '60ies:
the age of the by-pass, as an attempt to cope with the rapid growth in road traffic; at first with rather piecemeal by-passes, but very soon with an eye to an eventual motorway to be set up possibly by linking together various by-passes along a given route.
(iii) the fat years of the '60ies to the early '80ies:
the age of motorway construction on a big scale, along with major improvements to the national network.
(iv) since the late '70ies:
the age of regional development on a national scale: together with greater concern for safety and for the preservation of the environment.

All these 4 periods have implied a constant review of road schemes.

VI - SUCCESSIVE MASTER ROAD SCHEMES OVER THE LAST 25 YEARS

Roughly speaking, 5 milestones stand out inside the maze of successive road programmes that have been released over a quarter of a century: 1960, 1971, 1977, 1979 and 1984.

Each of these stages represents something of a Sisyphean task: each time you start from the network, which of course has progressed meanwhile, as it stands. Each time you take stock on the one hand of ever growing traffic and on the other hand of the overall funds to be expected. And each time, after long and rather painstaking studies (more for one means less for another), you are forced to carry out a comprehensive review of your overall master scheme. There is no point here in dwelling in detail on those 5 schemes, although the basic principles and guidelines may be of possible interest.

A swift glance at the two ends of the process may be of interest in this respect (See Figures 1 and 2).

(National network only)		1960 scheme	1985 scheme
(i)	Motorways	1,828 kms	6,520 kms
	Motorway extensions	1,478 kms	2,260 kms
(ii)	National roads 1st order	6,500 kms	6,830 kms
	2nd order	9,500 kms	
	out of a network of	81,000 kms	34,000 kms

Now, a few comments on that parallel:

(1) Just as a reminder, the decrease in the size of the national road network is a direct consequence of the redefinition of what the primary road network should be (see chapter II above).

(2) There has been a more than threefold increase in motorway lengths despite two main cuttings down to size after two reviews in 1977 (- 1570 kms) and in 1983 (- 1770 kms) which aimed at taking into consideration the new deal brought about by the crisis.

(3) A new factor should be noted in the 1985 scheme, namely that of "motorway extensions", by which I mean toll-free, high capacity highways (such as dual carriageways) extending the motorways and providing for the continuity of major primary

routes made up of the motorways and their extensions.

(4) As to the national road network itself, both schemes make provision for a sort of upper category of national roads. In the 1960 scheme, some 6,500 kms of "national roads of the first order" (chiefly three-lane carriageways felt to be adequate at the time); and in the 1985 scheme, 6,830 kms of "great links of national development". This is meant, in the views of the "Plan" and of "Datar" to complete, leaving no gaps, the structuring framework of the primary route network. These great links are thought to be closely adapted to the traffic volumes they are to accommodate.

(5) The 1960 scheme when released aroused a wave of criticism. Some found it disproportionate in relation to the actual needs of road traffic, above all considering the funds likely to be allocated and the limited number of personnel available in road departments (!...). Others rated that first scheme as a hopeless recurrence of the "dramatic error" of the railways in the XIXth century. A star-shaped motorway network centred on Paris was proposed, whereas the aim should obviously have been to dismantle that typically French, deep-rooted centralization - "un des grands desseins du septennat" - one of the grand presidential schemes of the present seven-year term. And the last straw: the example of the route from Paris to Bordeaux: half the stretch is motorway up to Poitiers; the rest, a conventional road. No idea of continuity and coherence!... It should be added here that the 1960 scheme had essentially been drawn up as a matter of expediency, under the pressure of public opinion eager for motorways, the issue of the day at that time. And the approach has simply been an extrapolation of the 1955 traffic census map of France, with a certain threshold above which a motorway was rated as necessary and below which a conventional road was thought to suffice. Hence that dead end at Poitiers for instance.

(6) No wonder then, in such conditions, that we had to start again from scratch. Things were taken up again as soon as the mid-'60ies. In short, the great difference between the approach of 1960 and that of later times lies in the generalization of the cost-benefit studies, a method so current in both our countries that there is no point in elaborating on it here. Imported from the USA at the end of the '50ies, those CBM (Cost Benefit Method) and PPBS (Planning, Programming and Budgeting System) of America were reshaped to meet conditions prevailing in France. Very rapidly, they became the open sesame of any scheme study, to such an

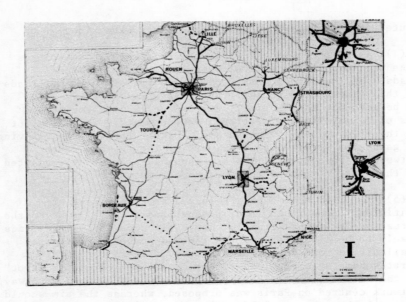

First such master scheme in France, released in 1960

Motorway programme:

— 1st urgency (before 1975) :   1,828 km
--- 2nd urgency :   571 km
.... 3rd urgency: 907 km
                     _____
                     3,306 km

                     rounded up to   3,300 km

National road programmes (of modernization)

— 1st order roads: 6,500 km
— 2nd order roads: 9,500 km

Fig. 1. The 1960 National Road Master Scheme

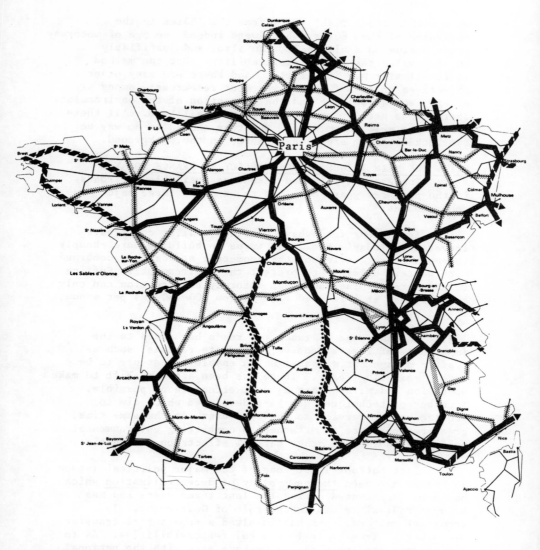

Motorways: 6,520 km
Motorway extensions: 2,260 km
Great road links 6,830 km for national development plans
National roads belonging to the national road scheme.

Fig. 2. The 1985 Master Scheme for the Primary Route Network of France

extent that those "fat" years from the '60ies to the beginning of the '80ies, which were indeed the age of motorway construction on a big scale, can also, and justifiably so, be called the age of profitability. But the method in itself also has its limits; and there are many other objectives to be met from Government resources: among others, social welfare and the protection of the environment. There is no need to go into further detail here. All these issues are well documented. But they should in no way be overlooked: they are always there, in the background.

VII - THE MAIN TRENDS OF ROAD POLICY IN FRANCE TODAY

Increasingly limited resources available for roads in the past few years have led to at least three major consequences:

1. Investment funds have declined, whilst costs have increased. Today's challenge turns on building more cheaply (hence a recent review of the geometric standards), combined with better building (to protect the environment). Suffice to add here for instance that motorway construction can only be continued at a reduced rate (below 100-130 kms per annum, at the present time).

2. The shortage of funds for roads has added to the importance of less expensive, immediate remedies such as traffic management measures which enable better use to be made of existing roads. There has been a clear shift to make operation and maintenance as cost-effective as possible. A number of techniques has been developed which aim to provide better service to the road user. At the same time, greater importance is attached to combatting environmental pollution, noise and other harmful effects on communities.

3. Last but not least, one of the major political issues in France has been the new policy of decentralization which has been implemented during the last three years and has brought radical reform to the role of Government. In transport matters, this has involved a substantial transfer of authority from central to local responsibilities. As to the primary road network, as we have seen with the national road system, this has meant financial innovation. Contracts between state and region are drawn up whereby both Government and regional councils commit themselves for the 5-year duration of the plan to pooling financial resources necessary for national road investments rated by both sides as deserving priority. Unlike the former system known as that of "contributed funds" (see Chapter III - paragraph A above),

when Government was the only decision-maker and was therefore solely responsible for decision-making and launching initiatives, "plan-contracts" (Contrats de Plan) as they are called in French, introduced a radically new method of co-financing road works which have of common accord been recognized as having priority in the national road system of the regions. Thus local authorities now have their say in national road programmes without in any way undermining the principle of the Government's leadership. Perhaps none of this surprises the British reader, but it must be clearly stated here, this would have been inconceivable in France in the former times of centralization which date back for centuries. And changes are not always simple to bring about (old habits die hard.....even in the civil service).

As to the expenditure that has to be met within the framework of such plan contracts, as a rule state and region each contribute half in rural areas, whereas in urban areas, where local authorities are more directly concerned, Government's average share drops to 27.5 per cent.

Finally, and to cut the story short, it must be understood that such projects are confined to by-passes or ring roads and that there will no longer be any Government aid for radials or cross-town links. And it must be clear too that from now on, the objectives more than ever are (i) better traffic flow, (ii) improved safety and (iii) improvement in the quality of living conditions.

VIII - CONCLUSION

In the present difficult, economic climate, the least that can be said is that the outlook for road construction in France today is rather bleak. The long and the short of it is that roads are no longer the major political issue they used to be. There is a clearcut shift in favour of public transport, along with a motorway network now more than 80 per cent completed, and a national road system modernized over two thirds of the network. The best that can be hoped is that things will continue as they are at present and that the 1985 national road master scheme will not have to be reviewed once again, nor cut once again. At the present time, this seems rather unlikely. However, a downgrading in the annual rate of construction can by no means be ruled out.

One final point: the Channel link. If that fixed link eventually takes shape (which now appears likely); a review of the main route schemes for the northern part of France will probably ensue with a view to either speeding up motorway construction throughout the area or complementing the existing network with main roads serving the coastal areas of Normandy and Brittany. But that's another story.....

6. Electronics for dynamic traffic management: the Dutch connection

J. J. KLIJNHOUT, Head, Electronic Systems, Rykswaterstaat

INTRODUCTION

When compared with the way industry plans the most economical use of its plants, the national motorway network is hardly managed at all. Every single driver tries to find his own best way through the network, a forlorn post mounted matrix sign might produce a weary smile. Occasionally a new stretch of road is opened to the public, but the glory of the opening is soon shaded by the dull experience that, also on these sections, weather and traffic conditions make driving a burden. Once finished the M25 around London, too expensive to be called just a ring road, it is said will be Britain's most expensive parking place. What drivers, and fleetowners in particularly have found long ago is still underestimated by most Authorities: delays on the major road network are frequent, costing and not very predictable. This article might give you some idea of what dynamic traffic management can mean to traffic on motorway networks. It is based on the experience we got in the past 5 years in the Netherlands with our Motorway Control and Signalling Systems (MCSS). To give away the cue: Motorway management systems have to be complete if they should be successful, successful systems are not cheap but worth the money.

ROAD CONSTRUCTION AND TRAFFIC MANAGEMENT

There is no doubt that some of the new motorway sections offer quite spectacular examples of modern civil engineering. Although not a beauty from environmental point of view, acceptable solutions have been found to fit the motorway in the landscape. The costs are on average millions of pounds per mile. Going metric might help to let it look cheaper. Scared by these costs two major faults are made:
-1- Construction is delayed to save money and plans are

OPERATIONAL ASPECTS

made more for the final futuristic situation than for the long intermediate period. Whatsoever the value of planning programs to find the best layout of the network, if the network plans are not implemented on schedule, a replanning is necessary to find the best optimum, since conditions will change with time. And even with a flexible road construction program certainly for as long as the final plans have not been implemented completely, the existing network will not be suited to handle the traffic it has to handle. Sit and wait for better times, doing nothing to make life easier is then the first fault.
-2- The second fault is that many expect that any new section of motorway is made to last for many years, and thus there should be no problems handling traffic: signallingcosts are considered too high.

An often abused argument in these situations is that the motorways are so expensive. It should be just the opposite, the argument should be used to justify the fractional increase in costs to install signalling to allow a better use of that expensive infrastructure.

The cost-benefit analysis carried out in the Netherlands showed that very broadly speaking full scale motorway signalling is justified on roads with peak hour loads over 1800 veh./lane.hour] when there is a zero increase in flow per year. These figures are indicative only. Detailed calculations can be made per country, per site, per situation. And in situations like network rerouting, or in typical adverse weather situations etc. break even point is reached earlier. The evaluation has shown us that although the reduction of the number of accidents is a psychologically easily explanable most quoted reason for signalling, the reduction in delays brings in about half of the revenues.
The experience with the Dutch system has also shown that the efficiency in economic terms, the credibility and thus the effect on traffic as well as the usefulness or applicability depends very much on the quality of the traffic responsive programs and on the resolution of (the distance between) the signs. Not on the number of bells and buzzers that are used to impress and convince the drivers, or on the amount of effort put into enforcement.
From the reactions of police officers visiting our MCSS site I know that the latter sounds to good to be true, but for the positive results with our MCSS, we needed no police enforcement and advisory maximum speeds are used. In all these years the police have had only one "testcase", bringing a driver who had ignored the

advised maximum speed to court. Their policy is that the system should sell itself, if not the system is wrong.

MCSS

Before saying more about the benefits of MCSS it might be good for those who have not yet seen to give a short description of it.

MCSS is operational since 1982. Detectors are of the inductive loop type. Double loops are used to get accurate (speed) data and to allow a check against malfunctioning. There are detector sites every 500 meters with double loops in each lane. Each site has one microprocessor which scans the loops, checks the data and transmits data (and status information) to any (up to 3) upstream outstation not more than 1500m away. The microprocessor automatically tunes the loops. Outstations are placed alongside the road near the gantries, the secretsigns of which they control. Also these outstation have a microprocessor to:
process the detector data of downstream detector sites, to transmit the results on request to a central computer and to control and monitor the secret signs. The secret signs are placed, one per lane, over the lanes. They are of the fibre-optic type, as big as the standard UK matrix sign but with a much better readability. They can show in white 30, 50, 60, 70, 80 or 90, a left or right pointing, slanting, arrow or, in red a cross. A computer centre covers up to 375 of such outstations. It will collect every 4 seconds traffic data, and every 20 seconds system monitoring data like the legends actually shown. Traffic data is handled by the centre's AID algorithm which will, when there is a queue, fully automatically immediately implement a slow down sequence of 70's and 50's. The gantries (and outstations) are spaced not closer than sign readability distance (300 m) and not further apart than retention time (1 min.) i.e. not more than 1500 m.
Data transmission from detectorstations to outstations and between outstations and control centre is via standard but dedicated (copper) transmission cables. Apart from its use as fully automatic AID system, the system can also be used by motorway maintenance personnel and by the police for lane closures and speed measures.

OPERATIONAL ASPECTS

MEASURES IN MCSS

The word measure is used here for the combination or sequence of legends shown on the signs. Some typical examples are AID measures upstream of and in a queue, and lane reservations.

- AID upstream the queue.

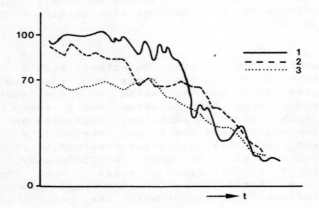

Fig.1. Traffic speeds when approaching a queue.

Accident statistics have shown that a protection of a queue by a police car on the hard shoulder, "flagging traffic down" guards sufficiently against so called secondary accidents by cars running into the queue. Measurements showed that in such circumstances traffic speed drops to some 70 km/h and vehicles come to a standstill after, smoothly. Without this advance police warning deceleration is more abrupt. Therefore in MCSS as soon as a queue is detected 50's will be shown on the gantry immediately upstream this queue, and 70's on the one "before" that. This mechanism works, speeds drop to 70-80km/h at the 70 showing gantry, the number of secondary accidents is reduced by 40 -50%.
- Measures in a queue.

As long as there is a queue the 50 will stay. When a queue grows and comes nearer to the 70 showing outstation, a new measure will be implemented, giving a 50 there and a 70 lead in upstream. Thus a queue is always followed by the system with a series of separate measures. This will lead to a succession of 50's over a long queue. The effect on traffic is both number and severity of shockwaves in the stop and go traffic in that queue are lower than without signing. The effect on capacity is an increase in throughput by 5%.

-Lane reservations.

When is lane is to be closed to traffic, two successive gantries are used. The first shows a slanting arrow over the lane which should be left accompanied by 90's over the other "open" lanes. The second shows a red cross and 70's. For sites with longer lane closures every gantry in that site will show the same X - 70 combination. The highway code forbids explicitly the use of the crossed-off lane. The arrow is just advisory. Still more than 95% of the traffic on that lane has moved over to an open lane before passing that arrow. Merging goes more smooth than with any other lane reservation arrangement ever used, we find that we can handle 1750 veh/hr instead of the 1100-1350 we were used to. In any planned lane reservation a trailer with a "keep right" or "keep left" traffic sign is placed, and alongside the reserved lanes cones should make clear that even though one has passed a maintenance crew the lane is still not open. Extra safeguards to take no unnecessary risks. With emergency cases like accidents the mere gantry signs proved sufficient.

The legends used in measures are carefully selected to tell the driver precisely what to do, and to in the case of an arrow-X sequence give him a second chance when missing a sign might be fatal. Secondly every care is taken to ascertain that signs are used only when necessary. The reason why is explained soon enough then by the circumstances.

As a result MCSS has become a very effective tool to manage motorway traffic. A somewhat negative example of the need to signal in an acceptable way: The effect of 70's alongside a crossed-off lane is that speeds go down from 100-120km/h to 80-90km/h. Better than in some long term road reconstruction situations where we have measured speed profiles which hardly differed from free flow traffic profiles, in spite of the fact that a complete arrangement of stationary mandatory signs including max. speed 70 km/h was placed. The strict use of signalling for lane reservation only when actually needed by the repair crews is an MCSS asset. Still the 70 signs with lane reservation are not as effective as the "same" 70 used in queueprotection measures.

CAPACITY

Rather than giving more statistics about the effect of MCSS, we give some conclusions of interest to those who want to manage the traffic on their motorway networks.
1. Delays caused by roadworks, accidents and bottlenecks are much greater than we expected. Many of these

OPERATIONAL ASPECTS

Motorway A13 on a snowy afternoon, with a lane reservation measure to protect men repairing a broken-down wide lorry standing on the hard shoulder. Note that cones etc. are not being used. The absence of tracks in the snow tells a tale of effectiveness.
Copyright: New Civil Engineer

delays are caused by temporary roadworks, and their total effect was underestimated.
2. The facilities to set up and remove lane reservations in a minute proved to have various positive effects,
   2a. As more traffic can be handled there is less risk for queues.
   2b. When a partially closed carriageway cannot handle all traffic, throughput is still 5% better than without signalling.
   2c. Lanes are closed to traffic during a shorter period. So when there is a queue caused by these lane reservation measures it will be there for a shorter time. Queues take time to dissolve, queues block traffic and make thus that queues grow. This cumulative effect makes it so important that the cause, a lane reservation is removed as quickly as possible.
3. Measures under oversaturation conditions to smooth traffic flows with advisory speeds, so as to prevent traffic coming to a standstill because of shockwaves can increase capacity by 4-5%. But there is no reliable automatic program yet for such actions.
4. Especially on heavy trafficked roads a close automatic monitoring is necessary to react rapidly whenever things go wrong. Thus
   4a. AID can reduce not only the number of accidents and the throughput problems caused by these accident, also 4b. the time it takes before the police is informed about an accident is reduced from 10-20 minutes to a few, with a faster clearing of the road as result.
   4c. The facilities the police have for lane reservations make life much simpler for them. The effect that has on capacity has not been quantified.
5. In more general terms, traffic flows form such a dynamic process that any management of them requires a high resolution in surveillance.
6. Although there are up to now no sites suited for ramp-metering to avoid oversaturation problems, the experience with MCSS shows that such surveillance is necessary for a good ramp-metering. Only in very rare cases of a comparatively very severe bottleneck might take the risk and ignore the effect of weather and of accidents and apply ramp-metering.

REROUTING

Rerouting signing is standardized in the Netherlands. At decision points in black on orange extra direction information is given with text saying which rerouting-codenumber to follow for which destination.

OPERATIONAL ASPECTS

This colour combination is standard for temporary text messages as well as for temporary rerouting information. All along the alternative route the rerouting-codenumber is shown. This system functions well. It is included now in MCSS where for rerouting secret signs of mechanical type (metal blinds) are used. These give problems and we are looking for better solutions, maybe fibre-optic or flip-disc versions.

Up till now rerouting is used only when it is definitely advantageous to all traffic rerouted.

IN CAR ELECTRONICS

Various road-vehicle communication systems are tested out in various countries. International cooperation should lead to an European standard. The value of these functions depends fully on the quality of the information about the road network situation. The electronic tools will function technically, there is no doubt about that. What however will be difficult is, to ensure that the responsible authorities know what happens out on their roads, and know what advice to give. To make these new electronics a success they will need monitoring systems of a standard more or less equal to that of MCSS.

CONCLUSIONS

This article does not tell how to solve traffic management problems. Instead I have tried to make clear that since we started to use MCSS we have learned a lot about the dynamics of traffic. We learned how much delay can be avoided by MCSS. Effects which give a better operational use of the network. We tried to avoid commanding drivers when we were not certain about the need. We tried to create credibility by given crisp and correct advise. We feel that although this limits what we can do, it certainly is worth doing. Expanding our management task by taking in rerouting, ramp-metering tidal-flow etc. is now in progress. It is will be done with an upgrade of the existing MCSS. Because we believe that although some prefer simpler systems, traffic dynamics and credibility would make that false economics.

Discussion on Papers 5–6

MR W. BRADSHAW, British Railways Board
With reference to Papers 5 and 6, I would like to make the following comments. It is heartening for a former Director of Operations of British Rail to see slides of a motorway signalling scheme which has so many characteristics of railway signalling. The French Railways have for some time been using advisory indicators in the complicated network approaching Paris. Drivers take pride in adhering to displayed speeds and this indicates to me that drivers of road vehicles would respond to a system from which they derived obvious benefit. Do the Dutch use closed circuit television to assist in 'spotting' accidents and have they incorporated any fog warning devices?
  May I ask Mr Thiebault whether he sees any future for toll motorways in France?

MR A. WHITFIELD, Department of the Environment and Transport – West Midlands Region
I would not like our overseas speakers to go away thinking that the Department of Transport was not prepared to try alternative means of traffic management.
  However, on Midlands roads, especially on the M6, we are having to handle very large volumes of traffic daily. We are looking carefully at all means of improving flow capacity and comfort.
  In April 1986, we hope to begin the first trial use of ramp metering as a means of sophisticated access control.

MR K. RUSSAM, Transport and Road Research Laboratory
As a researcher, I obviously want to see improvements made, but am aware, as we all are, that they have to be paid for. Our Dutch colleagues, in recently installing a signalling system on part of their motorway network, have been able to use the latest technology with impressive improvements in safety and operational effectiveness, for which they are to be congratulated. It was partly for this reason that the Demonstration Project of the EUCO-COST 30 international co-

operative exercise, in which the United Kingdom took a significant role, was sited in Holland.

In the United Kingdom, most of the motorway system has long been fitted with a communication system which, although somewhat dated, has helped to sustain an enviable safety record. The problem now is to update the system while continuing to optimise the cost/benefit equation; marginal improvements might be possible but at too high a price. However, the work is well in hand: last year, plans were announced of a major trial, on 80 km of the M1 motorway, of automatic incident detection equipment coupled with more closely spaced signals, about 1 km apart. The results of this trial will affect future provisions. It is the primary aim of the trial to compare, over a period of a year or so, the accident figures of the enhanced northbound carriageway with those of the unchanged southbound carriageway, and to establish whether there has been a reduction. Unfortunately, too many varying factors preclude accurate comparisons with other sections of the motorway.

MR THIEBAULT
In reply to Mr Bradshaw. Motorway financing by way of tolls implies and requires collecting enough money from traffic to meet all financial commitments. However, with the increase in construction costs (which means borrowing more money) and with the rise in loan rates (which means having to pay back more), this has become a difficult undertaking. With the present state of affairs in France, if the building of a new stretch of toll motorway is being considered - this new stretch being taken as such and in isolation, that is, without being bailed out by the profits that may be made from the existing toll network - there has to be a projected average daily traffic of 20,000 vehicles/km, if only to make both ends meet, financially speaking. In France, we have had the bitter experience of new toll motorway stretches that did not pay, chiefly because of the extent of traffic evasion, which had been underrated; believe me, such a situation is a real predicament. Eventually, we were rescued by the time factor, when traffic had at last increased sufficiently to bring in enough money from the tolls. I would simply say that the future for toll motorways in France is dependent on the profits than can be made from tolls. At the present time, few possibilities remain for new toll motorways in France. For this reason, our system of new toll motorway concessions appears to be tailing off, from a financial point of view. The point at issue is whether we will find another solution for the problem of how to finance new motorways; that is the question that for the moment remains unanswered. The state budget allows for more motorway construction, but only at a snail's pace.

DISCUSSION

MR KLIJNHOUT
In reply to Mr Bradshaw. Reliance is placed on the Automatic Incident Detection (AID) algorithm to "spot" accidents. Television is used only at those sites where it is desirable to know as soon as possible what kind of emergency services are needed in the case of an accident; which means, in practice, only in and around tunnels.

Fog detectors have been tested; however, although some of these function satisfactorily, for a good coverage they have to be placed every kilometer along both sides of the motorway. There are no so-called fog prone black spots, such as valleys, in our motorway system. Our experience is that AID effectively prevents multiple collision type accidents by warning traffic in time against crawling or standing traffic ahead. This all made us decide against the expensive installation of fog detectors.

In reply to Mr Whitfield. Ramp metering is possible in our MCSS because we have foreseen the use of this facility; however, the site has to have certain characteristics for ramp metering to be successfully applied. Also, the experiences, like those on the QEW in Toronto and in Washington, show that quite sophisticated surveillance is needed to ensure that some real benefits will result from metering. Research is continuing in order to gain a better insight into traffic behaviour, such that our models can correctly identify what degree of metering to apply. In this context, I would like to refer to the work of Payne and Papageorgiou.

I wish Mr Russam every success with his trial, but would, most respectfully, recommend him to review the scheme of comparing the effect of a well-equipped northbound carriageway with the unchanged southbound carriageway. Our experience has been that traffic behaviour during different times of day and under different circumstances (morning peak - evening peak) differs a great deal, probably too much for such a comparison to be valid.

7. Major missing links — the user's view

R. K. TURNER, MSc, BSc, MICE, MIHT, Director of Planning and Information, Freight Transport Association Ltd

SYNOPSIS. Industry depends fundamentally on road freight transport as an integral part of its production and distribution processes. Vehicle, network, and management developments over the years have been well used to reduce unit costs and increase efficiency. But there are still many gaps in the road network: physical gaps, overloaded sections, and sections suffering from major reconstruction when full of traffic. Industry depends on roads for a reliable and predictable level of service - unpredictability greatly reduces the industrial benefits of roads far beyond the simple effects of the local problem. Evaluating the real benefits of roads and facilities to users is a difficult business which has not yet been mastered and there are special considerations relating to the use of lorries to be taken into account when deciding whether or not to build a road as a single or dual carriageway.

INTRODUCTION
1. In 1984 we all spent, one way or another, over £20 billion on freight transport by road or about £20 per week for every household - the significance of that expenditure speaks for itself. The efficiency of that expenditure is dictated by competition, the vehicles themselves, and the quality of the network. This paper is primarily concerned with the quality of the road network but before turning to that issue it is important to be satisfied that industry, and particularly here, the road freight industry, is efficient and exploiting the network well.

THE EFFICIENCY OF THE INDUSTRY
2. In the last 10 years the work done by all lorries moving freight has increased by 19 per cent from 89.9 to 106.9 billion tonne-kilometres, but users expenditure (or the cost) over the same period has declined by 22 per cent, representing a total improvement of cost effectiveness of 34 per cent. These figures are illustrated in table 1.

Table 1   **Changes in freight transport demand and cost**

	1974	1984
Freight moved by road billion tonne-kilometres (all vehicles)	89.9	106.9
Users expenditure, road freight £billion, 1984 prices	25.9	20.1
Expenditure per tonne/kilometre, £	0.29	0.19

Source: Transport Statistics 1984

Table 2 shows how haulage rates have moved, emphasising how the intense competition in the road freight sector has effectively reduced rates by 27 per cent in real terms over the last 10 years.

Table 2   **Comparison of haulage rates 1974-1984**

20 tonne consignment ex London area for Birmingham 1984 prices.

	1974	1979	1984
£ per tonne	9.53	8.27	6.81(-27%)

Source: FTA Cost & Rates Service 1974-1984

3. Table 3 looks at the longer history of lorry weights, dimensions, motorway mileages and efficiency, illustrating dramatically how industry as a whole has used and taken advantage of the benefits of better vehicles, roads, and operating practices.

Table 3   Changes in lorry rules and numbers

	1955	1964	1968	1981	1983	1984
Maximum weight (tonnes)	24.3	32.5	32.5	32.5	38.0	38.0
Maximum length (Metres)	11.0	13.0	15.0	15.0	15.5	15.5
Nos. over 1.5 tonnes ULW (thousands)	530	627	637	498	492	482
Total tonne kilometres (billions)	37.6	61.6	78.9	87.0	91.0	96.6
Work done per lorry (thousands)	71	98	124	175	185	200
Motorways open, miles	0	292	563	1639	1693	1745

Source: Transport Statistics, Armitage Report, and FTA estimates.

HOW INDUSTRY WORKS

4. Missing links or gaps in the network are not just the obvious discontinuities of blue or green line on national maps. Of course, the completion of the M20 at Ashford is vital if the M20 is to mean anything to industry. Equally we need to join our major exporting and manufacturing centres to their prime markets and export links - the ports. But there are many missing links for industry which are not so obvious. To understand where they are it is helpful to understand the way industry works and how the road network is used.

5. Crudely, 85 per cent of all freight transport goes by road but even more important than that - road transport of freight is the only universal common denominator. Every factory, warehouse and customer, has a road service but few can have a direct service by rail, water, canal or pipeline. Therefore the vast majority of journeys start and finish by road even if the intermediate operation is by another mode. The road vehicles themselves come in a wide range of shapes and sizes but in recent years more and more vehicles have become specialised for one particular use. To quote tanker vehicles is an obvious example. Less obvious are the vehicles specialising in hanging garments, palletised loads of building materials, or chilled food. With this specialisation of lorries has come increased capital cost - a large high pressure tanker costs up to £100,000 to buy. A more typical lorry will cost about £18,000 to buy and between £12 and £14 an hour to run with the largest lorries costing between £24 and £32 an hour.

6. Increased cost in the vehicles themselves is often matched by increased costs from specialised handling (for example frozen food), warehousing, stockholding, and the high cost of financing a large inventory (goods in the "pipeline"). Freight transport therefore is an integral part of a manufacturing, warehousing, distribution and marketing process. It is not an end in itself but an increasingly sophisticated link in the production - customer chain. As such, therefore, it has to fit in with the demands and needs of other parts of that chain. Vehicles arrive at warehouses according to pre-determined schedules - an idle vehicle in a queue waiting to be unloaded is inefficient, and it is often crucial that certain deliveries are received first and in order. For example, a department store would want to take fresh food in at the beginning of the day and reserve unloading space accordingly. Clothes and other non-food items are not so urgent and would be unwelcome early in the morning.

7. It is clear, therefore, that as far as industrial road transport is concerned, timing, scheduling and predictability are most important. Similar arguments would apply to scheduled passenger services by road but would not apply so accurately to other passenger vehicles, particularly those on recreational journeys. Of course, predictability is important to all road users for road safety reasons - an unexpected delay will encourage faster driving to make up lost time. Nevertheless, for industry, the benefits of good roads can only be fully achieved if they are predictably and reliably better.

8. Consider the example of a journey between London and Birmingham. In theory a lorry can easily make that journey in 2 hours at an average speed of 50 mph. However, if delays on the M1 through congestion or road works mean the journey time varies between 1.8 hours (55 mph) and 2.5 hours (40 mph) the transport manager would be forced to plan for the worst time, or else labour, equipment and other vehicles might be held up because of this vehicle's late arrival. Therefore, in this theoretical (though realistic) example, the economic benefits of the motorway are devalued by 20 per cent (50 mph to 40 mph) <u>all the time</u> because of <u>unreliability</u>.

9. The sensitivity of industrial transport planning to predictability is further increased by legislation restricting lorry drivers to an eight hour driving day. Therefore, as well as the obvious organisational penalties of a vehicle not arriving on time, any schedule which uses the whole eight hours at a high optimum level of driving performance, carries an extra penalty when delays occur - the vehicle and driver are stranded away from base or their scheduled overnight resting or parking place.

THE GAPS IN THE NETWORK

10. It can readily be appreciated, therefore, that there are three types of gaps in the network:-

a. the obvious gaps like M20, A30 Okehampton bypass;

b. gaps created by overloading and unreliability like Dartford Tunnel, M63 Barton High Level Bridge; and

c. gaps created by the lack of capacity to handle major reconstruction like M5, M1 and M6.

11. Last year the CBI* published an analysis of many of the gaps in the first two categories and the BRF published** two reports detailing the need for more investment in roads On the 8 November 1985, the CBI published an update called "Fabric of the Nation Mk II". In particular it called for a fresh look at the COBA assessment method (following FTA criticism of its under-valuation of industrial benefits during the NEDC review earlier in 1985), and a proper appreciation of the dynamic benefits of road investment (the glow around the wire). The appendix to this paper compares the "Fabric of the Nation" schemes with the "1985 White Paper Roads Review". It shows in summary form the industrial need, region by region, it however omits a number of important needs off the trunk road network, for example a third tunnel at Dartford.

12. Of the 44 'The Fabric of the Nation' schemes involved, all bar three are dealt with to some degree by the 1985 Review.

13. However, this does not mean that all of the schemes in 'The Fabric of the Nation' are to be fully implemented. In aggregate, the D.Tp schemes in 'The Fabric of the Nation' are estimated to cost about £3,750 million (November 1983 prices). At present, the D.Tp projects which make up these schemes have been allocated around £2,500 million. If the cost of the three schemes not included in the programme(3) (£500 to £750 million) is deducted, <u>a difference of £500 to £750 million still remains between D.Tp planned spending and that which was called for in 'The Fabric of the Nation'</u>.

---

*Fabric of the Nation, 1984. Confederation of British Industry.
**Challenge and Opportunity, 1984. British Roads Federation.

MAJOR MISSING LINKS

14. This leaves the third category of gap - those created by lack of capacity to handle major maintenance and renovation. It is clear now that our major highway network should to some degree carry redundant links or capacity which can provide alternative routes and extra needed space when major works are under way. This was dramatically illustrated in July 1985 when junction 8 of the M1 was reconstructed. A reduction in traffic capacity of 50 per cent was forced into the network for 2 weeks and there was no alternative route. A completed M40 could have provided that alternative. Equally, we have now to accept the difficulties of designing roads for both a 20 year traffic capacity and 20 year structural life - they wear out just when they are full and reconstruction will be most difficult.

15. There can be no universal solution to this problem - the extravagance of doubling trunk road capacity is no answer. But in some cases, on vital links, extra lane capacity earlier than would be justified by normal traffic growth must be provided. Equally, as illustrated by the M40 example, extra capacity in the network elsewhere will often provide a less ideal but acceptable alternative route for much traffic, making reconstruction possible without dramatic reductions in the levels of service or predictability for industry.

OTHER FACTORS

16. There are other factors which introduce gaps or poor performance into the road network - particularly for industry.

17. **FACILITIES** - this is probably the most obvious. Europe's largest bypass, the M25, will be complete at the end of 1986 at a total cost near to £1,000 million, yet this 116 miles of high grade motorway will not have one toilet, one telephone, one resting place, or one fuelling point. Apart from the obvious safety risks of drivers going beyond their natural endurance or breakdowns through lack of fuel, there is the extra disbenefit of vehicles going into London in search of these facilities, the very place nobody wants them and they do not want to be. The sad fact is that nobody disagrees with this analysis but we are told our planning system is such that the D.Tp are impotent to speed up the processes. Two, three or four years after the road is open service areas will be built. But have we got our priorities right?

18. **RESTRICTIONS** – the primary route network is by definition the best route between primary destinations and as such lorries and their drivers are encouraged to use it in preference to other routes. The primary route network, therefore, is no place for weight restrictions, height restrictions, or other impediments to industrial traffic.

19. **DOUBLE DISBENEFIT** – where restrictions do occur it is often because the road is inadequate environmentally to accommodate the traffic and in need of improvement. But, if the best route between primary destinations is restricted, industry suffers a double disbenefit: the lack of a better route, which an improvement would have provided, plus the extra penalty of being banned onto a less direct route.

20. **CONFIDENCE** – industry invests in warehousing for a life of 5 to 10 years and factories for a life of 15 to 20 years. These planning horizons are much longer than the short term planning of industry which looks just 2 or 3 years ahead. For the longer plan to be successful industry needs confidence in the planning environment surrounding its investment. Road planning uncertainty currently exists for anything up to 15 or 20 years – this uncertainty will be reflected in industrial planning decisions.

THE EVALUATION OF BENEFITS

21. The FTA has long held the view that the benefits to industry of an improved road network are under-valued by the COBA* analysis system. This system relies on assessing the time savings of vehicles using the new link (or route) as compared with the old route. In effect this provides an economic assessment of the simple benefits of a link rather than the contribution of that link to the network. The industrial benefits of roads are only real when time savings aggregate to real vehicle savings similar to the bus lane theory. While it is possible to evaluate that a bus lane saves say 40 passengers 15 seconds each, the value of that saving to the economy is very dubious. However, lots of bus lanes saving lots of bus time can be shown eventually to save buses with fewer needed to do the same job – a real saving.

---

* The cost benefit analysis system used by the Department of Transport to evaluate the economics of building a new piece of road.

22. The same applies to lorries, but even more so because time savings on a network can and do result in structural changes in distribution systems, with fewer depots and distribution and stockholding in completely different places and organised in more cost effective ways. A test study by the Lorries and the Environment Committee* showed that for one national drinks company total warehousing and distribution costs could be reduced by 6.4% following the 16 year motorway programme finishing with the M25. In this case depot numbers would be reduced from 20 to 17. Other companies have already achieved even greater savings. A number now operate from one depot only in the Midlands serving the whole country by road reducing stockholding and improving levels of service. Others in the food sector have reduced from 20 or more depots to just 5 or 6, one each main centre of population.

23. It is these structural effects which are so dramatic and valuable that the current COBA method takes no account of. But this is not an ignorant omission, because these savings are so difficult to evaluate and economists prefer to use time savings as a convenient proxy for the structural cost savings. FTA is not confident that this proxy is sufficient and further research is under way.

24. But before leaving COBA it is worth recording that the weaknesses in its technique are of little import if the Government are purely to use it to prioritise schemes in a preparation pool much bigger than the resources allocated to it. If COBA is to be used to the full effect it must be taken into account when the Treasury allocate public expenditure funds. Only then is it worth making sure that it is refined to reflect the full industrial benefit. The 1984 National Road Traffic Forecasts give even more weight to the argument that there is little merit in high precision allocation of grossly inadequate funds.

FILLING THE GAP

25. The way in which a missing link or gap in the network is filled often can in itself turn out to be controversial. At the time of writing this paper the A30 from Exeter to Redruth was a very good example. Not only does this route pose extremely controversial questions about the actual route around Okehampton (issues which incidentally are frustrating to the user - we just need a good road somewhere) but long standing arguments about the quality of the route: should the A30 be developed as a dual or single carriageway?

---

*A Pilot Project to Assess the Benefits of the Road Building Programme to one company - an unpublished report by the Lorries and the Environment Committee dated July 1981.

26. First there is the obvious economic argument about spending our resources wisely. Dual carriageways are more expensive than single carriageways (3 times more expensive if one accepts the Parliamentary Written Answer on 21 March 1985*). Assuming a finite limit on road building monies, there has to be overwhelming traffic and economic argument to justify the extra expense. Second, and in contrast to the above, there is the economic and operational argument from the road user - and particularly the industrial user. Dual carriageways crucially have different speed limits for lorries, see table 4.

Table 4   Lorry Speed Limits

	Speed Limit	Typical Scheduled Speed	Time for 250 miles	Distance in 8 hr day
Single carriageway	40mph	25mph	10.00hrs	200miles
Dual carriageway	50mph	35mph	7.14hrs	280miles
Motorways	60mph	45mph	5.55hrs	360miles

Source: FTA

27. All lorry drivers are limited by EEC Regulation 543/69 to an eight hour driving day. In theory, therefore, a driver can cover 320 miles on a single carriageway de-restricted road. In practice that is not possible and most companies have agreements with their drivers and their unions on the performance to be expected on typical roads. These agreed scheduled speeds are built into the working timetable of the operation so that whole distribution systems can operate reliably. Timing and scheduling are important to the industrial process to avoid queues, over-stocking, or "stock-outs". Therefore, returning to table 4, it can be seen that a 250 mile trip is well outside the capabilities of a lorry on a single carriageway but well inside its capabilities on a dual carriageway.

28. With the use of computers it is possible to be far more sophisticated than these simple calculations. DAF Trucks, for example, have a well researched system of analysis which was designed to help potential customers choose the right truck for a particular operation. It enables any specific vehicle to be run, by computer simulation, over any combination of a 1000 kilometres of European roads. The output provides fuel consumption, gear

---

*Dual 2 lane all-purpose road £3.6m per mile, single carriageway, £1.2m including an allowance for land, rehousing and supervision.

MAJOR MISSING LINKS

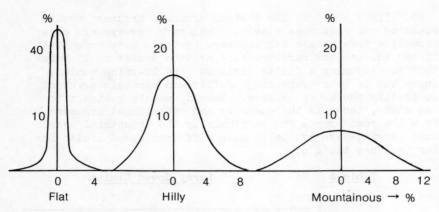

Figure 1. Road characteristics of DAF TOPEC simulation programme

Source: DAF Trucks Eindhoven, Holland

Table 5   A30 Exeter to Redruth planned and preferred improvements

	Miles	Current	Planned	Preferred
Exeter to Okehampton	5.4	SC	D2	
Okehampton Bypass	5.5	SC	D2	
Okehampton to Launceston	12.2	SC	SC	D2
Launceston to Plusha	5.9	SC	SC	D2
Plusha to Bolventer	8.7	SC	SC	D2
Bolventer to Bodmin BP	2.0	D2	–	
	4.5	SC	–	D2
Bodmin Bypass	6.3	D2	–	
Bodmin Bypass to Indian Queens	7.0	SC	–	
Indian Queens Improvement	3.8	SC	D2	
Summercourt Improvement	1.7	SC	D2	
Mitchell Bypass	1.6	SC	D2	
Zelah Improvement	2.0	SC	SC	
Zelah to Blackwater	6.0	SC	–	
Blackwater Improvement	2.2	SC	SC	
Blackwater to Camborne	1.5	SC	–	
Camborne to Redruth Bypass	6.0	D2	–	
Total single carriageway		68.0	53.8	22.5 mls
Total dual 2 lane c'way		14.3	28.5	59.8 mls
	82.3	82.3	82.3	82.3 mls

SC - single carriageway
D2 - dual 1 lane carriageway
Source: British Road Federation and FTA

changes and speed. However, by fixing the vehicle specification and performance and varying the route characteristics, the effects of different road standards can be compared with some confidence. Figure 1 shows the three versions of route available varying from flat to mountainous. The latter was discarded as unrepresentative of the A30 and in the main, the "hilly" route was chosen. Tables 5 and 6 summarise the results of the work.

29. Table 6 shows the marked reduction in trip time resulting from increased investment in dual carriageway layouts. These benefits are further enhanced in areas remote from the main markets by increasing assessability within the 8 hour driving day. Therefore, the 1.18 hours saving on the preferred design as compared with the existing route will increase the theoretical daily travel horizon by 54.5 miles (1.18 x 46.2).

CONCLUSION

30. Industry has a fundamental and indepensible reliance on roads as an integral part of the manufacturing and distribution process. The full benefit from road investment depends on the quality of the original construction, plus

Table 6. **Simulated lorry journey A30 Exeter to Redruth Bypass.**

	Current		Planned		Preferred		All Dual
	SC	D2	SC	D2	SC	D2	D2
Av speed mph	23.3	46.2	30.4	46.2	30.4	46.2	46.2
Av consumption litres/100km	55.8	37.0	32.1	37.0	32.1	37.0	37.0
Trip time hours	2.91	0.30	1.77	0.62	0.74	1.29	1.78
Total trip time	3.21		2.39		2.03		1.78
Gear shifts	352	16	70	33	52	69	95
Revs per KM	2272	1185	1572	1185	1572	1185	1185

Assumptions: 38 tonne lorry fully laden, 4.0m high, aerodynamic drag coefficient 0.80, rolling resistance 6.50 kg per tonne, hilly route conditions (see fig.1).

Source: FTA based on DAF TOPEC analysis system.

MAJOR MISSING LINKS

the proper provision of essential services and efficient maintenance. Gaps in the network are all capacity related Capacity to carry the traffic, capacity to provide services needed by users, and capacity to cope with essential repairs.

31. Industry needs a greater commitment to the provision of reliable road systems justified by an assessment of the benefits of roads to businesses and maintained to a high standard of reliability to optimise these benefits.

PAPER 7: TURNER

APPENDIX. COMPARISON BETWEEN 'FABRIC OF THE NATION' AND THE '1985 ROADS' REVIEW'

Fabric of the Nation scheme (cost £m)[1]	DTp projects	DTp Implementation date	Cost £ million Nov. 1983 prices	Date in Fabric of the Nation	CBI Comments[2]
Greater London					
A406 East London River Crossing (99/130)	East London River Crossing	1989 onwards	130.0	1987-89	Little scope for acceleration. Public enquiry not yet held.
A40 Junction improvements (55/53)	Ealing Gypsy Corner	1987-89	15.7	1987-89	Inefficient to have more than 2 schemes in progress together.
	Ealing Western Circus	1987-89	15.2		
	Long Lane junction	1989 onwards	15.0		
	Swakeleys Road junction	1989 onwards	7.2		
A12 Eastway-Eastern Avenue (88/85)	A12 Hackney Wick to M11 Link	1987-89	85.0	1987-89	
A406 North Circular Road Improvement (333/368)	Links under construction	1985-87	60.7	(1985-87 (1987-89 (1989 onwards	Traffic requirements preclude more schemes before 1989.
	Great Cambridge road (A10) junction	1985-87	21.4		
	Popes Lane to Western Avenue	1987-89	21.3		
	Hanger Lane to Harrow Road	1987-89	50.9		
	Falloden Way junction	1987-89	24.2		
	Chingford Road to Hale End Road	1987-89	42.7		
	Regent's Park road junction	1989 onwards	41.0		
	Golders Green junction improvement	1989 onwards	10.0		New in programme.
	Bounds Green/Green Lanes improvement	1989 onwards	40.0		
	East of Silver Street to A1010	1989 onwards	23.4		
	Dysons Road to Hall Lane	1989 onwards	32.9		
A23 Improvement (52/4)	Coulsdon Inner Relief Road	1989 onwards	4.0	1991-94	
M25 widening (150/-)	-	-	-	1991-94	Not in programme study in progress.
South East					
A23 Crawley-Brighton Improvement (20/28)	Warninglid to Brighton	1987-89	24.0	1987-89	New in programme (not in Fabric of the Nation).
	Handcross to Pease Pottage improvement	1989 onwards	3.7		
A34 Oxford-Winchester Improvement (19/28)	East Ilsley to Chilton improvement	1985-87	3.5	1987-89	
	Whitway diversion	1985-87	5.2		
	Newbury Bypass	1987-89	19.5		
A41 Improvements in Herts and Bucks (34/25)	Kings Langley Bypass	1987-89	10.4	(1985-87 (1987-89	DT83: 1985/87.
	Berkhampstead Bypass	1987-89	10.3		
	Aston Clinton Bypass	1987-89	5.0		DT83: 1987/89 reserve list.

The first figure refers to the estimated cost of the scheme in 'The Fabric of the Nation', the second to the amount allocated in the 1985 Roads Review. All figures are in November 1983 prices.

DT83 = Status of road project in the 1983 Roads Review.

Table continued on pages 110-113

MAJOR MISSING LINKS

Fabric of the Nation scheme (Cost £m)[1]	DTp projects	DTp Implementation date	Cost £ million Nov. 1983 prices	Date in Fabric of the Nation	CBI Comments[2]
South East (cont)					
10 A3 Improvements in Surrey and Hants (38/43)	Campton-Shacklefield improvement	1985-87	6.0	(1985-87 (1987-89	
	Midford Bypass	1987-89	4.3		
	Liphook and Petersfield Bypass	1987-89	23.7		
	Hindhead improvement	1989 onwards	11.8		DT83: 1987 onwards
11 A27/A259 Havant-Folkestone improvement (160/147)	Havant to Chichester Bypass	1985-87	29.3	(1985-87 (1987-89 (1989-91	
	Fontwell improvement	1985-87	3.6		
	Westhampnett Bypass	1987-89	6.5		
	Brighton Bypass	1987-89	52.8		DT83: 1985-87.
	Pevensey Bypass	1987-89	4.2		
	Winchelsea Bypass	1987-89	3.1		
	Arundel Bypass	1989 onwards	11.8		New to programme.
	Patching Junction improvement	1989 onwards	5.0		New to programme.
	Polegate Bypass	1989 onwards	8.3		New to programme.
	Bexhill and Hastings Western Bypass	1989 onwards	16.2		New to programme.
	Rye Bypass	1989 onwards	5.8		
	New Romney Bypass	-			Suspended.
	Dymchurch and St Marys Bay Bypass	-			Suspended.
	Hythe Bypass	-			Suspended.
12 A420 Swindon-Oxford improvement (28/2)	Kingston Bagpuize and Southmoor Bypass	1987-89	2.0	(1987-89 (1989-91	Need to add further schemes. Public consultation underway.
13 A2 Dualling in East Kent and future upgrading to motorway (105/10)	Barham Crossroads junction	1985-87	1.8	1991-94	Need to upgrade Section NW of Dover No upgrading to M-planned.
	London Boundary to M2 (provision of hard shoulders)	1985-87	8.1		
Eastern					
14 A11/A47 Newmarket-Great Yarmouth inc Norwich Southern Bypass (111/77)	Thetford Bypass	1985-87	6.6	(1985-87 (1987-89 (1989-91	
	Wymondham-Cringleford	1985-87	8.8		
	Newmarket-Red Lodge	1987-89	1.0		
	Red Lodge Bypass	1987-89	1.9		
	Acle Bypass	1987-89	4.6		
	Besthorpe-Wymondham	1987-89	3.0		
	Norwich Southern Bypass	1987-89	44.2		
	Four Wentways junction	1987-89	2.9		
	Roundham Heath to Snetterton dualling	1989 onwards	1.8		New to programme.
	Four Wentways-Newmarket dualling	1989 onwards	1.2		
15 A418 Thame-Stevenage Route (105/-)	Thame to Stevenage	-		(1989-91 (1991-94	Suspended.
16 A47 improvement King's Lynn to A1 (23/19)	Peterborough Westwood Junction	1985-87	1.2	(1985-87 (1987-89	DT83: reserve list 1987 onwards.
	Peterborough Longthorp Junction	1985-87	1.2		
	Eye Bypass	1985-87	3.8		DT83: 1985/87 res list.
	Guyhirn Diversion	1985-87	2.8		DT83: 1987 onwards
	Caster and Ailsworth Bypass	1987-89	2.8		DT83: 1985/87 res list.
	Thorney Bypass	1987-89	2.4		DT83: 1985/87 res list.
	Walpole and St Johns	1987-89	2.9		DT83: 1987 onwards Possible need for further dualling.
	Tilney High End Bypass	1987-89	1.8		
17 A12/A47 Great Yarmouth Bypass completion (6/6)	Great Yarmouth Southern Bypass (Southern section)	1987-89	5.9	1987-89	Other sections completed or under construction.

PAPER 7: TURNER

abric of the ation scheme Cost £m)[1]	DTp projects	DTp Implementation date	Cost £ million Nov. 1983 prices	Date in Fabric of the Nation	CBI Comments[2]
outh Western					
A361 North Devon Link completion (42/58)	Barnstable Bypass	1985-87	14.8	(1985-87	
	North Devon Link stage 2A	1985-87	25.0	(1987-89	
	North Devon Link stage 2B	1985-87	18.0		
A417/A419 improvement (15/13)	Birdlip Bypass	1985-87	1.5	(1985-87	
	Blunsdon to Cricklade widening	1985-87	2.1	(1989-91	
	Brockworth Bypass	1987-89	6.8		DT83: 1985/87.
	Latton Bypass	1989 onwards	2.5		New in programme.
	Stratton Bypass	-			Suspended.
A30 Exeter-Penzance improvement (56/88)	Blackwater improvement	1985-87	4.3	(1985-87	
	Okehampton Bypass	1985-87	22.0	(1987-89	
	Exeter to Okehampton stage 3	1985-87	19.2		
	Zelah improvement	1987-89	3.1		DT83: 1985/87.
	Mitchell Bypass	1987-89	2.1		DT83: 1985/87.
	Summercourt improvement	1987-89	2.2		DT83: 1987/87.
	Indian Queens improvement	1987-89	5.8		
	Plusha to Bolventor	1987-89	8.7		
	Lauceston to Plusha	1987-89	6.2		
	Okehampton to Launceston	1989 onwards	14.4		
A36 Bristol-Southampton Road improvement (48/33)	Bath to A420	1985-87	3.5	(1985-87	
	Warminster Bypass	1985-87	9.0	(1987-89	
	Heytesbury Bypass	1985-87	1.6	(1991-94	DT83: 1985/87 reserve list.
	Steeple Langford Bypass	1985-87	1.5		DT83: 1985/87 reserve list.
	Batheaston and Swainswick Bypass	-			Under study
	Beckington Bypass	1987-89	1.6		DT83: Reserve list 1987 onwards.
	Cod Ford Bypass	1987-89	1.5		DT83: 1985/87 reserve list.
	Salisbury Bypass	1989 onwards	14.0		
A40 Oxford-Ross-on-Wye improvement (74/35)	Highnam to Jays Green improvement	1989 onwards	5.5	1989-91	
	Witney Bypass to Cassington dualling	1989 onwards	6.2		New in programme.
	North of Oxford improvement	1989 onwards	23.5		New in programme.
Second Severn Crossing (105/-)	-	-	-	1991-94	Not in programme but under study.
st Midlands					
A564 Stoke-Derby link (91/89)	Stoke-Derby link (Hatton, Hilton and Foston Bypasses)	1987-89	9.8	(1985-87 (1987-89 (1989-91	DT83: 1985/87.
	Stoke-Derby link (Stoke Southern Bypass)	1989 onwards	17.0		DT83: 1987 onwards.
	Stoke-Derby link (Doveridge Bypass)	1989 onwards	7.6		
	Stoke-Derby link (Derby Bypass-Derby Spur)	1989 onwards	54.2		
A446 Birmingham Northern Relief Route (116/120)	A446 Northern Relief Route	1989 onwards	120.0	(1987-89 (1989-91	Little scope for acceleration.
M40 Oxford-Birmingham (226/223)	Oxford-Birmingham (Otmoor section)	1987-89	40.8	(1985-87 (1987-89	All four - DT83: 1985/87. No acceleration possible. Schedule likely to be delayed as not feasible to start all four schemes together.
	Oxford-Birmingham (Gaydon section)	1987-89	25.5		
	Oxford-Birmingham (Warwick section)	1987-89	46.0		
	Banbury Bypass	1987-89	111.0		

111

MAJOR MISSING LINKS

Fabric of the Nation scheme (Cost £m)[1]	DTp projects	DTp Implementation date	Cost £ million Nov. 1983 prices	Date in Fabric of the Nation	CBI Comments[2]
West Midlands (cont)					
27 A435 improvement - (M42-M5 link) (23/50)	Alcester Bypasses	1985-87	7.4	(1985-87	
	Evesham Bypass	1985-87	9.5	(1987-89	
	Studely Bypass	1987-89	13.8		
	West of Ashchurch - A435	1989 onwards	6.7		
	Norton and Lenchwick Bypass	1989 onwards	12.7		
28 A449 Birmingham West Orbital Route (79/-)	Wolverhampton Western Bypass	-		1989-91	Suspended.
	Stourbridge Bypass	-			Suspended.
	Kidderminster Eastern Bypass	-			Suspended.
29 M6 duplication in Cheshire/Staffordshire (394/-)		-		1991-94 1994 onwards	Under study but not in programme.
East Midlands					
30 A604 M1-A1 link (86/102)	A604 M1-A1 link (Kettering-Brampton)	1985-87	44.4	(1985-87 (1987-89	Danger of slipping back to 1987/89.
	A604 M1-A1 link (Kettering section)	1985-87	13.8		
	A1-M1 link (M1 - Kettering)	1985-87	44.1		
31 M42/A42 completion	Measham-Ashby Bypass	1985-87	23.5	(1985-87	
	Castle Donington North	1987-89	24.1	(1987-89	
32 A43 Oxford-Stamford improvement (51/50)	Peartree Hill to Wendlebury improvement	1985-87	20.6	(1985-87 (1987-89	Little scope for acceleration.
	Brackley Bypass	1985-87	6.0	(1989-91	DT83: reserve list 1985/87.
	Towcester Bypass	1985-87	4.2		
	Bulwick Bypass	1985-87	2.3		
	Kettering Northern Bypass	1985-87	5.0		
	Blisworth Bypass	1987-89	8.2		
	Silverstone Bypass	1989 onwards	3.4		New in programme.
	Whitfield Turn to Brackley Hatch dualling	1989 onwards	2.5		New in programme.
	Stamford Bypass	1989 onwards	2.5		DT83: reserve list 1987 onwards.
33 A46 Leicester Westerm Bypass (33/34)	Leicester Western Bypass	1987-89	34.3	1987-89	
34 A6 Improvement in Leicestershire and Northants (32/43)	Market Harborough Bypass	1987-89	5.0	(1987-89	
	Barton Bypass	1987-89	2.3	(1989-91	
	Great Glen Bypass	1989 onwards	1.9		New in programme.
	Burton Latimer to Rushden	1989 onwards	13.6		
	Clapham Bypass	1989 onwards	4.5		New in programme.
	Kegworth Bypass	1989 onwards	2.1		
	Quorn and Mountsorrel Bypass	1989 onwards	12.7		
35 A47 Leicester-A1 dualling (80/4)	Billesdon Bypass	1985-87	2.6	(1985-87 (1987-89 (1989-91	Need to get more schemes into programme.
	Wardley Hill improvement	1985-87	1.6		DT83: 1987 onward
North Western					
36 M63/M66 Manchester Outer Ring Road East Flank (165/180)	Portwood to Denton	1985-87	57.0	(1985-87	Little scope for acceleration.
	Denton to Middleton	1987-89	122.0	(1987-89	
37 M63 Stratford-Eccles improvement (11/20)	Stratford to Eccles improvement			1985-87	No scope for acceleration. DT83: 1983/85.
	Stage 1	1985/87	8.3		
	Stage 2	1985/87	2.4		
	Stage 3	1985/87	2.5		
	Stage 4	1985/87	6.6		

PAPER 7: TURNER

Fabric of the Nation scheme (Cost £m)[1]	DTp projects	DTp Implementation date	Cost £ million Nov. 1983 prices	Date in Fabric of the Nation	CBI Comments[2]
North Western (cont)					
38 A590-M6 to Barrow development (15/12)	Dalton in Furness Bypass	1987-89	7.7	(1987-89 (1989-91	
	High and Low Newton Bypass	1989 onwards	3.0		New in programme.
	Swarthmoor Bypass	1989 onwards	1.2		New in programme.
39 M65 Calder Valley Route completion (53/50)	Blackburn-M6 link	1989 onwards	50.0	1991-94	
40 A59 Liverpool-Preston improvement (89/9)	Burscough Bypass	1987-89	9.1	(1985-97 (1987-89 (1989-91	
	Bickerstaffe to Brotherton	-			Suspended.
	Bretherton to Hutton	-			Suspended.
	Preston Southern Bypass	-			Suspended.
Northern					
41 A69 improvement (68/65)	Horsley to Corbridge improvement	1985-87	2.4	(1985-87 (1987-89 (1989-91	DT83: 1985/87.
	Eighton Lodge Junction improvement	1985-87	4.0		
	Newcastle Western Bypass	1985-87	49.0		
	Brampton Bypass	1987-89	3.9		DT83: 1985/87.
	Warwick Bridge Bypass	1989 onwards	5.1		
	Haltwhistle Bypass	-			Suspended.
	Haydon Bridge Bypass	-			Suspended.
42 M1/A1/A6183 Kirkhangate-Dishforth (127/130)	Wetherby Bypass	1985-87	9.3	(1985-87 (1987-89	
	Kirkhamgate-Dishforth (Knowsthorpe-Austhorpe)	1987-89	13.2		DT83: 1985/87.
	(Austhorpe to A1 Bramham)	1987-89	21.7		DT83: 1985/87.
	(Stourton to Knowsthorpe)	1987-89	9.0		DT83: 1985/87.
	(Lofthouse to Stourton)	1987-89	13.3		DT83: 1985/87.
	(Dishforth Interchange)	1987-89	4.6		DT83: 1985/87.
	(Bramham-Wetherby)	1987-89	37.5		DT83: 1985/87.
	(Wetherby-Dishforth)	1987-89	9.3		DT83: 1985/87.
43 A629/A650 Airedale Route (89/65)	Airedale route (Victoria Park to Crossflats)	1985-87	10.5	(1985-87 (1987-89	
	Airedale route (Kildwick to Beechcliffe)	1985-87	19.5		
	Airedale route (Crossflats to Cottingley Bar)	(1987-89 (1989 onwards	2.2 17.0		
	Airedale route (Shipley Eastern Bypass)	1989 onwards	15.5		
44 A616/A628 Sheffield/Manchester Route (79/18)	Stocksbridge to M1 Mottram to Tintwistle Bypass	1985-87	17.5	(1989-91 (1991-94	Suspended.

113

8. Major missing links — examples from the South East

J. S. DAWSWELL, MBE, BSc(Eng), FICE, Chief Engineer, Transportation Engineering Division, W. S. Atkins & Partners

SYNOPSIS. The transportation division of my firm has been involved with some examples of missing links in the highway system in the South East of England. This paper looks at some of the details and problems of four such schemes. The schemes are: The M20 Motorway between Maidstone and Ashford in Kent; The A13 between the M25 and Rainham in Essex; The A12 Hackney Wick to the M11 Link Road; and the A406 South Woodford to Barking Relief Road, the later pair being in North East London. The paper attempts to draw some conclusions in connection with the delays which have caused these links to be missing from the current road network.

INTRODUCTION

1. A motorist visitor to this country deciding on suitable routes for a journey might well be struck by the apparent gaps in the blue lines on maps which represent our motorway system, and in some cases be even more surprised when the gaps are not filled by a thick red line representing a reasonably high quality trunk road. At first sight it might be difficult to understand how our major network has been allowed to develop to its present state when the gaps involved are in many instances without prospect of being filled within the next five years. The reasons are of course many and varied and, as will be seen from this paper, result in part from our UK determination to be democratic in our own particular way, taking all points of view into account before making any major change to anything. This style differs somewhat from our European colleagues. The cost to the country which results from this must be enormous although somewhat indirect. Perhaps we should consider whether or not the price is worth paying.

THE M20

2. The M20 is an important route to the Channel ports of Folkestone and Dover. The only gap in the blue line is between Maidstone and Ashford. In this instance there is a trunk road available but obviously it is not of the standard offered by the motorway sections. The provision of a new

motorway between the eastern end of the Maidstone Bypass and Ashford was investigated in the 1960's and possible interchanges between the M20 and A20 were part of this study. The preliminary report on the scheme was submitted by the Kent sub-unit of SERCU in 1969.

3. Discussion on the report and further investigation resulted in a change of preference to a full interchange at Hollingborne, for which side and connecting road orders were published in 1972. A public inquiry in 1973 resulted in an Inspector's report in 1974 which, although finding in favour of the proposals generally, recommended alterations to the line of the motorway over the western half of the scheme. This necessitated the withdrawal of the interchange proposals and the withdrawal of the side road proposals in this area. In 1975 the Secretary of State, while accepting the recommendations of the Inspector, announced his decision to confirm the main line from Lenham Heath to Ashford and to defer consideration to the section between the Maidstone Bypass and Lenham Heath. In early 1979 the modified main line order was published in draft and a Public Inquiry held at the end of the year. Feasibility drawings were displayed to the public in connection with the interchange proposals, based on those previously published in 1972. The Inspector noted the representations made at the Inquiry and commented that he found no reason to disagree with the proposal for a full interchange at Hollingborne.

4. Following receipt of the Inspector's report, further studies were again made into the interchange options. The conclusion was to keep the full interchange at Hollingborne. In June 1980 the entire Maidstone to Ashford scheme was temporarily shelved, probably due to financial priorities, and taken out from the list of Schemes in Preparation. The final section of main line was confirmed in 1981.

5. During 1982 the scheme was reconsidered and included in the list in the 1983 White Paper.

6. In Autumn 1983, W S Atkins & Partners were commissioned to progress the scheme and this has been such as to work towards publication of the Side Road Orders, Connecting Road Orders and Compulsory Purchase Orders at the end of 1985. An advance contract for a railway bridge was started in the summer of 1985 and another advance contract for a bridge to take the A20 trunk road over the line of the new motorway will go to tender at the beginning of 1986.

THE A13

7. The A13 has for a long time been an important trunk road in the network, serving as a vital link between the Docks at Tilbury and the centre of London. Its importance

was reinforced in the late 70's when a new, very large container terminal was built on reclaimed land at Tilbury.

8. In the early 70's proposals for improvement to the trunk road between Tilbury and Rainham were finalised and taken to a Public Inquiry in 1972. This involved relocation of the A13 on an entirely new route. The inspector reported favourably on the proposal but it was decided to go ahead with construction from the east only as far as the line of the then proposed M25 motorway. This decision was not taken because the section 'inside' the M25 was unimportant but because it was thought that to construct it would prejudice discussions and decisions on that part of the route to be taken through or round Rainham.

9. Proposals had already been formulated to improve the existing A13 along the existing route through Rainham, this being cheaper than other options. Very strong representations had been made against the route through Rainham and various bypass routes to the south of the town through the marshes were proposed. The Department of Transport had not favoured routes through the marshes on geotechnical and other grounds, and the consequent cost. The construction cost was estimated to be much greater and the distance travelled by vehicles would also be increased, compared with the on-line route through Rainham. Depending on which route was selected for the Rainham area, the route between the M25 and Rainham might need modification from that proposed for its westernmost 'half' which had been planned to join the existing A13 a little to the east of Rainham.

10. Thus in the late 70's construction commenced on the section of the new A13 to the east of the M25. The interchange at the M25 (junction 30 of the M25) was constructed so that the A13 could easily be continued westwards whatever decision was finally made for the Rainham area. This eastern section of the A13 together with the M25 is of course now open to traffic, but the driver arriving at the M25 junction 30 from the Tilbury area after travelling along the new A13 will be relatively surprised to see the A13 flyover bridge pointing westwards into space while he is directed southwards into the probable traffic jam at the roundabout at the northern end of the Dartford Tunnel approach (part of the M25 junction 31) and then back on to the old single carriageway A13 road into London. A glance at his map would reveal an obvious location for the unbuilt section of the A13 between the M25 and Rainham and the driver will be puzzled that it has not been constructed when there seems no obvious physical obstruction.

MAJOR MISSING LINKS

11. In the early 1980's the effects of completion of the M25 became more and more apparent as did the need, as a consequence, to construct this missing link of the A13. In view of the urgency a decision was made to improve the A13 on its present line through Rainham and, therefore, to build the missing link on the line which had been taken to Public Inquiry in 1972. While the various statutory processes were being organised and prepared, strong representations were made for reconsideration of the marshes route and in early 1984 the Minister for Transport announced that there would be a reappraisal of the various route options. This reappraisal showed that the cost penalties of the marshes route were perhaps now less than had been assessed some years before, when compared with other environmental factors. It was, therefore, decided to go to Public Consultation to obtain reactions from various Authorities, local inhabitants and road users, to four selected viable proposals. The results from this consultation are at present awaited, but meanwhile the link in question is still missing and will probably remain so until the next decade. I regard this as an example of the country paying a high price for our democracy.

THE HACKNEY TO THE M11 LINK ROAD

12. Perhaps one of the most expensive and most difficult to build of the missing links may be the Hackney to the M11 Link Road, officially called the London - Great Yarmouth Trunk Road (A12) (Eastway to Eastern Avenue Section). A glance at the map shows the M11 motorway bringing traffic from the North to the north-east side of London. It terminates at a roundabout on the A12 trunk road at Redbridge. Traffic wishing to continue in a southerly direction towards the river Thames has no obvious route, and traffic wishing to go radially towards the centre of London is faced with the A12, a dual two-lane road. Within half a mile of turning towards London, however, all hopes of continuing on a reasonably free highway disappear. One is at traffic lights in the middle of Wanstead or more likely, in a long queue leading up to the traffic lights, and the dual two-lane road reduces in width to single carriageway. Less than four miles away, further towards London, there is another stretch of blue on the map (A102M). This indicates in fact, a high quality highway from the north-east part of Hackney (at the river Lea) to the Blackwall Tunnel, with a grade separated interchange at the junction with the A11 at a point just over two miles from the City.

13. The journey between the southern end of the M11 and the A102(M) can be long and tedious at almost any time of day. The possible routes through Leytonstone and Leyton are more suited to traffic from the last century and one wonders that the need for the link has not already been satisfied by construction of a new highway or a radical improvement along existing routes.

14. The history of the proposed routes in this area is quite long and complex and of course has been related to various overall policies nationally and for the London area. After the Public Inquiry in 1959 an order was made in 1962 fixing the line for a new road between Temple Mills and the Green Man. This can be seen to be approximately the western half of the missing link under discussion. However, only a very small section of construction was implemented in Leytonstone near the Green Man roundabout. In 1973 there were public consultations on four possible routes between Hackney and the M11 and these included two proposals for a widening of the A12 between the M11 and the Green Man.

15. As a result of this, a preferred option between Hackney and the Green Man was announced in 1974. It was also decided to go ahead with the improvement scheme to Cambridge Park (ie. the A12 between the Green Man and the M11). It was stated at the time that further consideration would be given to improving the road system to the north of the Green Man towards the North Circular Road in conjunction with the proposal. The draft orders for the Cambridge Park improvement were withdrawn after they had attracted considerable objections.

16. The need for something to be done on this missing link became ever more apparent, and in early 1976 a Joint Working Party was set up to consider all aspects of the problem. This Working Party consisted of officers from both central and local government and it was hoped that all points of view would be adequately represented. The report of the Working Party, in considerable detail, was published in late 1979 and it included comments and assessments on quite a large number of options for each of the six sections of the link, together with a section related to possible road improvements to the north of the Green Man.

17. Responses to the reports were received and bearing this in mind, a preferred scheme was announced by the Secretary of State for the link road in the summer of 1981. The proposals for improving the existing highways north of the Green Man were ultimately dropped. This was, in part, finally made possible by the inclusion in the scheme of a much improved proposal for the road between the Green Man and the M11. A series of public exhibitions was held to explain the proposals and to invite comment. In particular, objections were very strong to the proposal to widen the existing road through the centre of Wanstead at existing ground level. This resulted in the inclusion of an expensive tunnel section at Wanstead which obviously improved considerably the route between the M11 and the Green Man. This tunnel is interesting from the engineering point of view as part of it will be just above the existing London Underground Central Line tunnels and underneath the existing A12 highway, which

will not be closed during construction. There were some 500 objections to the scheme and a Public Inquiry was held into the proposals between February and September 1983. It is interesting to note that at this Inquiry very few objections denied the need for the Link Road. By this time the traffic problems in the general area through which the link passed had become evident to all, and the general 'rat-running' by motorists in the area had become a pronounced nuisance. The London rush hour traveller who listens to the broadcasts on his radio telling of the traffic problems of the moment will seldom hear a transmission which does not include an item denoting a serious traffic problem somewhere along the line of Hackney to M11 Link Road. The main opposition to the proposal comes from those areas between the western end of the link road and the centre of London, and those who do not want any further intrusion into any land which is part of Epping Forest, a much prized amenity. The former rightly claim that whilst the new road may help the problems in the area through which it actually passes, it will do nothing for them and will only attract traffic on to their already congested and overloaded roads. For the most part the objectors along the line of the Link Road pressed for a more expensive scheme which would have put part or all of the new highway beneath ground level. Here perhaps I should mention a considerable contribution to the general discussion put forward by a local group of objectors and included as one of the options in the Joint Working Party report. This was for a full tunnel scheme with general improvement at ground level to include a linear park which would quite clearly have improved the environment in the area. After the proposal had not been accepted by the Secretary of State it was pursued through the Public Inquiry (as an objector's scheme) by the group led by a local architect who devoted a very great deal of his time and energies to promoting it. The Inspector of the Inquiry was clearly very impressed by the points made in favour of the scheme, but in the end its high cost for the benefits achieved prevented it being recommended. I think it fair to say that most members of the public nationwide (the taxpayers) would be unlikely to consider it the best way to spend an extra £100 million of public money. This is particularly so when there are so many other urgent needs for public expenditure, and resources far from infinite. As a result of the Public Inquiry and the Inspector's report, the Secretary of State announced in October 1985 the intention to go ahead with the scheme generally but with certain aspects modified to take into account the Inspector's recommendations. This included a revised interchange at the River Lea to help reduce traffic problems at the western end of the Link Road, and a tunnel beneath Epping Forest land at the Green Man interchange.

18. Most will be aware of this announcement which appeared in the national daily papers in October 1985, and indicated that the modified proposal would cost a total of some £125 million.

19. At the time of writing this paper, therefore, there is once again progress on the scheme but because of all the statutory processes through which the scheme still has to pass, it is unlikely that construction will begin much before the end of this decade and that the new route will be fully available to traffic before 1993. These processes include a Parliamentary Bill necessary for Epping Forest land to be used and other statutory processes must, therefore, be programmed around this.

THE A406 SOUTH WOODFORD TO BARKING RELIEF ROAD

20. The North Circular Road is a heavily trafficked London Orbital route extending from Chiswick in the west to Woolwich Ferry in the east, a total distance of about 45 km. It is a vital road for the movement of heavy commercial vehicles, in particular around the northern sector at London. From Chiswick to Waterworks Corner, the latter being just 1 km from the western end of the above Relief Road, the North Circular Road is a trunk road (A406). Much of its length is now dual carriageway and most of the remaining congested sections and junctions are going through the various statutory processes for improvement.

21. Between Waterworks Corner and the Woolwich Ferry the signposted North Circular Road is not a trunk road and there is a significant deterioration in its standard; it is not even easy for the motorist to find. Traffic is directed, or finds its own way, through both shopping and residential streets. The combination of this 'through' traffic, much of which is of heavy commercial vehicles, with local traffic, creates major traffic congestion and lengthy delays. Proposals for a relief route following the River Roding Valley to the A13 at Barking had been indicated for over 30 years on Development Plans. Originally known as the 'C' Ring Road, the title was changed to Ringway 2 in the draft Greater London Development Plan (GLDP) of 1969. It has also been known as the M15 but is now identified in the Department of Transport's trunk road programme as the South Woodford to Barking Relief Road.

22. The GLDP was approved in 1976 and whilst the plan does not include the original Ringway concept it was stated that "the North Circular Road, which is an existing orbital distributor in west and north London, will continue to be improved by the Department of Transport and extended to the A13".

MAJOR MISSING LINKS

23. I have already referred to the fact that traffic wishing to continue in a southerly direction from the end of the M11 has no obvious route.

24. Between 1973 and 1977 the southern section of the M11, terminating at the A12 Redbridge roundabout, was constructed. The layout of the M11 carriageways, at its southern end, allows for construction of the Relief Road and other possible then planned major highways such as the M12 and the M11 southern extension.

25. In 1976 members of the public were able to see, and to give their views on, outline proposals for the Relief Road. The route exhibited was similar to that already protected by the initial GLDP. In 1978 a Steering Group was set up, consisting of officers from both central and local government, to consider the incorporation of the route into the corridor in the context of national strategies and local planning, identifying options including feasibility, traffic, design, environment, and costs. The Report of the Steering Group was produced in 1980. Subsequently the Department published draft orders under the Highways Act, and the scheme went to a Public Inquiry in 1982.

26. Various alternatives to the basic route alignment were raised by objectors. In particular objectors favoured crossing beneath the British Rail tracks, in lieu of over, at Ilford, and also moving the Relief Road further west away from properties in Wanstead Park Road. All alternative proposals increased the scheme cost. As recommended by the Inquiry Inspector's report, the orders were made based on the Departments scheme with only minor modifications, and staged construction started in March 1985. The road is programmed to be opened to traffic in mid 1988 as far as the A13 at Barking. The link across the River Thames would be completed by the proposed East London River Crossing, which is currently at Public Inquiry.

COMMENTS AND CONCLUSIONS

27. The speed with which we progress our highway schemes is related to a number of factors. Priority in relation to need is perhaps the most important of these since the total yearly budget available for expenditure on road construction is limited. Only a certain proportion of national expenditure can be allocated to highway construction and improvement. Clearly we cannot build at a faster rate than we can afford. The amount of money that should be spent each year is a matter for our elected government to decide and not a subject for this paper.

28. Our legislation is complex and has been built up over centuries. At the present time we have a policy which gives most careful consideration to all aspects of highway

construction together with the effect on communities and their environment and also to an evaluation of the financial benefit to the country as a whole. As the years go by, the Department of Transport refines this latter in an attempt to compare as accurately as possible the benefits of schemes. This enables us to invest the tax-payers money in a way which will achieve most benefit.

29. We have statutory procedures which ensure that problems which may arise from new highway construction are considered during the planning stage, and we hope that all people who may be affected are adequately warned and able to make representations about their fears and concerns. It is, I think, this latter which is the biggest cause of delay to our highway schemes, some which have been desperately needed for a long time. This may be seen from some of the schemes I have discussed earlier and also the fact that the M25 is still not complete.

30. Causes of delay are often quite unfairly and wrongly attributed to the engineers responsible for a scheme, who are making every effort to progress new proposals and improve our trunk road network. That we should bring about delay by giving so much consideration to the views of every citizen is perhaps not in the interests of the vast majority of the population. We must, of course, accept that changes of Government policy and changes in the Government itself will result in a revision of priorities. This, combined with the changes in the economic fortunes of the country and changes in attitudes such things as visual intrusion and noise, will inevitably bring disruption and delay to new highway scheme preparation. However, where the line should be drawn is, I hope, a subject for discussion resulting from this paper.

9. The Ipswich bypass, western section — a case study

L. M. ELLISS, MA(Cantab), FICE, FIHT, Director, G. Maunsell & Partners

SYNOPSIS. The western section of Ipswich by-pass was the last gap in the otherwise dual carriageway standard route from the Midlands and the North (via A1) to the Haven ports. It was also the last link to be completed in the Ipswich by-pass system. The western section was opened some three years later than the adjoining southern section, both lengths having started the planning process together. The classic controversy between inner routes hugging urban development and outer routes through open countryside was the major cause of delay, resulting in two major public inquiries into the siting of the road.

BACKGROUND

1. <u>The Town</u>. Ipswich is an historic borough with a population of some 120,000. It is the county town of Suffolk, the largest town in the county and an expanding industrial, commercial and administrative centre. Ipswich is also a port, being situated on the navigable River Orwell about 15 km from the river mouth.

2. <u>The regional road network</u>. Ipswich stands at the intersection of two important trunk routes: the A12 from London to Great Yarmouth and the A45 from the Midlands to Felixstowe (Fig. 1).

3. The A12 London to Great Yarmouth trunk road is a radial route connecting Chelmsford, Colchester, Ipswich, Lowestoft and Great Yarmouth to London. It has been improved to dual carriageway standard from the outskirts of London to Chantry Park in the south-west of Ipswich except for a length at Chelmsford where a dual carriageway by-pass is currently being constructed.

4. The A45 Felixstowe to Weedon trunk road connects Felixstowe to the Midlands and provides a link between Ipswich, Bury St. Edmunds, Newmarket and Cambridge. There are continuous dual carriageways from the A1 (via A604) to Claydon to the north of Ipswich and from Ipswich to Felixstowe docks.

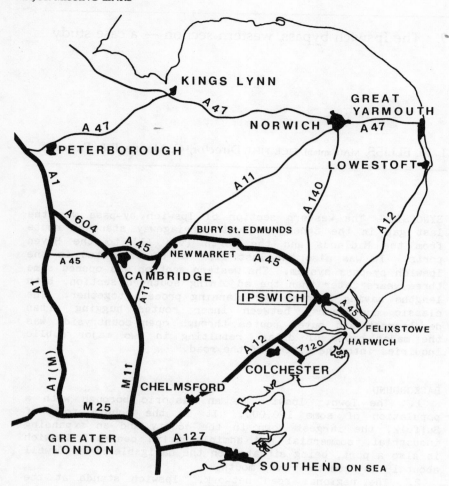

Fig. 1. Major road network in East Anglia

5. <u>The local road network.</u> While originally the A12 and A45 penetrated the centre of Ipswich, they had subsequently been united and diverted in an arc around the northern outskirts to form a joint A12/A45 ring road, constructed in the 1930s (Fig. 2). The ring road has built-up frontages, a single carriageway and numerous roundabout and traffic signal controlled intersections.

6. The former A1100 principal road (now B1113) runs southwards from the A45 at Claydon past the village of Bramford and through the village of Sproughton to join the A12 near Washbrook and Copdock. The A1100 was a primary route extensively used by traffic wishing to avoid roads

through Ipswich. The single carriageway road is generally of substandard width with poor vertical and horizontal alignment and acute pedestrian/vehicle conflict through Sproughton village, exacerbated by heavy sugar beet transporters requiring access to the nearby British Sugar Corporation works.

HISTORY OF THE PROPOSALS

7. <u>Government policy</u>. From the early 1970s central government has consistently given high priority to the construction of a by-pass system for Ipswich. During this period particular emphasis has been placed on the improvement of roads which aid economic recovery and development, specifically including links to major ports (e.g. ref. 1). Ipswich By-pass not only improves access to the ports of Ipswich and Felixstowe from both the Midlands and the South-East but in addition benefits traffic associated with the port of Harwich bound to or from the North. Between 1980 and 1984 the total cargo handled through the Haven ports (Felixstowe, Ipswich and Harwich) increased by 44% from 11.5m tonnes to 16.6m tonnes and looks well set to achieve the projected doubling of throughput in the decade.

8. Second only to this consideration has been environmental benefit, particularly the need to relieve communities suffering from the effects of lorries. Ipswich scores on this count also, being cited as a major example of a town choked by heavy traffic (e.g. ref. 2).

9. <u>Transportation and feasibility studies</u>. In 1970 a transportation study of Ipswich was sponsored by the Department of the Environment and Ipswich County Borough. Arising from this study a decision was made to investigate an outer by-pass of the town and a feasibility study of various routes north and south of Ipswich was carried out by East Suffolk County Council. Its conclusions indicated that a by-pass south of the town would yield the greater benefit to the community as a whole and on the 25th May 1972 the Parliamentary Under Secretary of State for the Transport Industries announced that a new by-pass, to be built around the southern perimeter of Ipswich, would be included in the trunk road preparation pool.

10. <u>Public consultation</u>. In 1974, in accordance with the decision of the Secretary of State for the Environment to give members of the public greater opportunity to participate in the early planning stages of road schemes, a public consultation exercise was held. Alternative routes were considered on each of the three sections of the by-pass to the west, south and east of Ipswich, including options for crossing the Orwell estuary by bridge or tunnel. A public exhibition at Ipswich attracted 3,500 visitors and more than 1400 people returned questionaires.

MAJOR MISSING LINKS

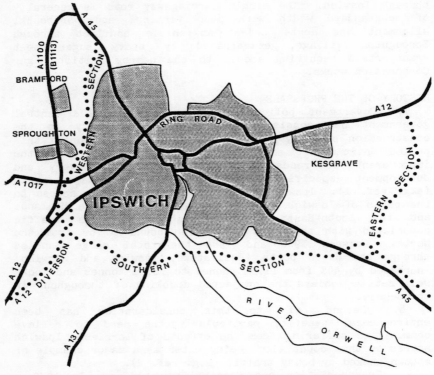

Fig. 2. Road Network around Ipswich

11. <u>Western section alternatives</u>. From the time of the public consultation exercise onwards two basic corridors have been considered for the western section of Ipswich by-pass (Fig. 3).

12. "Outer" routes pass through open, undulating farm land to the west of the A1100 (B1113) and the villages of Bramford and Sproughton. "Inner" routes are drawn closer to the existing urban development of Ipswich, separating Ipswich from Bramford and Sproughton and passing through an undistinguished and less actively cultivated agricultural area.

13. In the public consultation document the outer route is stated as being approximately 1km longer than the inner route (10km compared with 9 km) and is estimated to cost about £0.5m more at 1974 prices (£4.85m compared with £4.35m). The public response showed that the inner route was most popular but there was also considerable support for an outer route. Suffolk County Council, and Ipswich Borough preferred an outer route whilst Mid Suffolk and Babergh District Councils favoured the inner route.

14. **Preferred route**. In September 1975 the Department of The Environment announced that the Southern by-pass would cross over the River Orwell on a bridge. The western section would follow the outer route but would use the A1100 link road, then being constructed from the Claydon interchange as part of the improvement of A45 between Stowmarket and Claydon, rather than make a "head on" connection with the A45 as previously envisaged. (This strategy was designed to avoid duplication of nearly 1 km of dual carriageway including expensive structures over the River Gipping and the Norwich railway line). Survey and design work was then carried out and draft orders for the line of the by-pass were published under the Highways Acts in October 1976.

15. **The first public inquiry**. There were 128 objections to the draft proposals - the most significant for the western section of the by-pass being from Mr. N.J. Fiske, a local farmer, whose land was seriously affected by the published (outer) route. He proposed an alternative inner route, approximating to the inner route considered during the consultation exercise.

16. A public inquiry was held in Ipswich between May and October 1977 to consider all three sections of the by-pass, to the west, south and east of the town. The inquiry sat for 36 days having been disrupted at the beginning by procedural objections, which were common at that time.

17. The Department of Transport evaluated the alternative inner route and compared it with the published outer route. Traffic assignments predicted that 22% more vehicles would use the inner route in the design year (1997) than the outer route (25,400 per day compared with 20,800). The then current design standards (ref. 3) required grade separation where flows fell between 25,000 and 40,000 vehicles per day. For this reason the inner route was designed to be grade separated throughout whereas the outer route had generally at grade intersections. Although the inner route was some 600 m shorter than the outer route the additional expense of grade separation produced an estimated cost around £3 m higher (£9 m compared with £6 m). However, in view of the shorter route and the greater volume of traffic attracted to it, the economic indicator NPV/PVC was marginally higher for the inner route at 0.93, compared with 0.91 for the outer route. In addition the residual traffic on A1100 would be reduced by over 60% by the inner route compared with the outer route and Sproughton would also be relieved of the seasonal sugar beet traffic.

18. The inspector (Mr. F.H. Clinch) found that in addition to the matters referred to above, the inner alternative had the following advantages:-

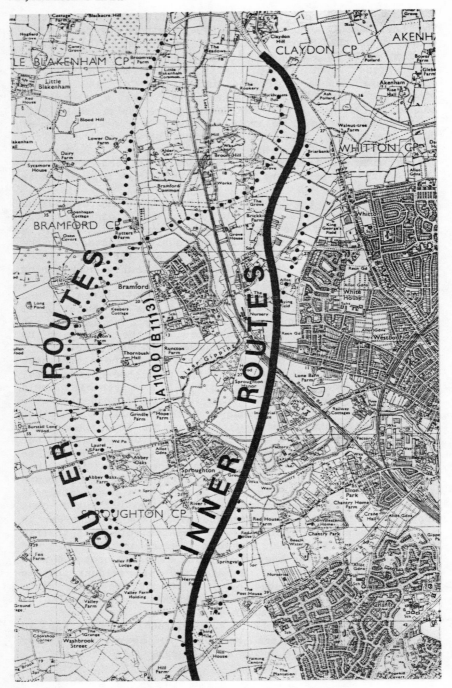

Fig. 3. Western Section alternatives

(a) It would serve the needs of Ipswich traffic better (particularly with regard to industry on the west side of the town) and bring greater traffic relief within the town.
(b) It had a superior horizontal alignment and the lead onto the by-pass at Claydon would be improved.
(c) Having full grade separation it would be a safer road and would have no impact upon movement within the surrounding area.
(d) It would not sever Sproughton and Bramford from their rural hinterland.
(e) The industrial estates at Whitehouse Road, Farthing Road and the sugar beet factory would all be better served.
(f) It would have a less damaging effect on the landscape and agriculture.

But it would have the following disadvantages:-
(a) It would have severe impact upon the plans of the Ipswich Borough Council for industrial and housing development.
(b) The noise nuisance at existing dwelling houses would be greater.

19. The inspector concluded that the merits of the published scheme and the inner alternative were nicely balanced. He considered that the additional cost of the alternative was justified as it would result in an improved economic performance and be more attractive for internal Ipswich trips. He therefore recommended that the inner route should be adopted.

20. On the 8th September 1978 The Secretary of State for Transport announced that he accepted the inspector's recommendations with regard to the western section of the by-pass and that new draft orders would be published, generally along the lines of the inner route considered at the inquiry. He said that the decision had not been easy but that he particularly had in mind the safety advantages of grade separated junctions which would be justified on a route closer to the town. When new proposals were put forward they would be designed to minimise the adverse impact on industrial and housing development.

21. After further survey and design work draft orders for the western section of the by-pass along the inner route were published in May 1981. In November 1981, after representations from Suffolk County Council, revised proposals were published for a full interchange with Sproughton Road rather than just the north facing slip roads previously proposed. Compulsory purchase orders for the western section, were published in draft in January 1982.

22. The second public inquiry. A total of 44 separate objections to the draft orders were received, the most substantial being from the Western Section Action Group,

representing some 1,000 people, which proposed an alternative outer route. 36 counter-objections to the alternative routes proposed by objectors were also received as well as a substantial number of letters in support of the published proposals.

23. Sir Alexander Waddell, an independent inspector nominated by the Lord Chancellor, conducted the second inquiry on ten sitting days in April 1982. He found that the alternative route proposed by the Action Group was effectively the published scheme considered at the 1977 inquiry except that the current proposal included two additional single carriageway link roads from the A1100 eastwards to Sproughton Road to the north and south of Sproughton. The object of the link roads was to relieve the village streets, particularly of industrial and sugar beet traffic. They would be the responsibility of Suffolk County Council (rather than the Department of Transport) to sponsor and fund.

24. As at the 1977 inquiry, an at grade outer route was compared with a grade separated inner route. Once again the traffic predicted for the inner route was considerably higher than for the outer route (Table 1) and the estimated cost was also greater (£21.5 m compared with £17.7 m, at 1982 prices). The cost advantage of the alternative route would, however, be reduced to about £3 m if additional survey costs and design fees were taken into account. At the 1982 inquiry the inner route was shown to have a significantly better economic performance that the outer route (rather than the marginally better return indicated at the 1977 inquiry - mainly due to developments in economic analysis). The published route was now given net present values of - £0.75 m and +£8.97 m for low and high traffic growth respectively and the outer alternative -£4.36 m and -£0.17 m.

Table 1. 1982 inquiry traffic figures.
(vehicles per day - high growth)

	Published (inner) route		Alternative (outer) route	
	1985	2000	1985	2000
Copdock - Sproughton	20,000	28,300		
Copdock - Bramford			12,200	18,300
Sproughton - Whitehouse	17,900	26,000		
Bramford - Claydon			15,800	23,500
Whitehouse - Claydon	32,200	46,500		

25. The inspector in his report examines the alternative routes from a number of aspects and, in addition to the matters referred to above, draws the following conclusions:-
   (a) The surface junctions and crossings for farm vehicles on the northern part of the outer alternative route would present hazards to all drivers.
   (b) The north and south link roads at Sproughton would provide greater relief for the village than the Department's route but there was no certainty that they would be built.
   (c) The outer route would inflict considerably greater damage on agriculture and the landscape than the inner route.
   (d) Noise increases from the Department's route affect considerably more properties than the alternative route.
   (e) Acceptance of the alternative route would cause a delay to construction which the Department put at 2 years 9 months. This is an important but not overriding factor.

The inspector concluded that the alternative (outer) route was not to be preferred to the published (inner) route and recommended that the orders should be made.

26. The Secretaries of State, in their decision letter of 18th April 1983, accepted that the alternative (outer) route, as a development of the route advocated by the Department of Transport at the 1977 inquiry, deserved careful consideration even though that route was subsequently rejected by the inspector and by them. They agree that delay alone should not be an overriding factor and that the outer route has some advantages but state that in a number of important aspects it is significantly inferior to the currently proposed (inner) route. They therefore agree with the inspector that it is not be be preferred to that route and confirm that the orders for the inner route will shortly be made.

27. <u>Construction</u>. The southern section of the by-pass (including the major bridge over the River Orwell) was completed in December 1982 followed by the eastern section and the A12 (Copdock to Washbrook) diversion in 1984. Tenders for the western section were invited from selected contractors in June 1983. The works comprised 8 km of dual two-lane and a similar length of single carriageways, two intermediate grade separated interchanges and eight structures. Earthworks consisted of about one million cubic metres of excavation, some two thirds of which is used for embankments. The contract was awarded to Balfour Beatty Construction Ltd. at a tender price of £15.3m (with the main carriageways to be in concrete) and work started in February 1984 with a two year construction period (Fig.4). Despite

Fig. 4. Construction between Sproughton (lower left), Bramford (upper left) and the western outskirts of Ipswich (right)

an exceptionally wet summer in 1985 construction was completed four months early with the road being opened to traffic on October 3rd, 1985.

CONCLUSIONS

28. Having started the planning process at the same time as the southern section, the western section of the by-pass took nearly three years longer to accomplish. In the intervening period not only had traffic been denied the use of the western section but conditions on the A1100 through Sproughton became even worse. Heavy goods vehicles, in particular, diverted from routes through Ipswich to gain access to the southern section.

29. It would be easy, with hindsight, to say that the Department should have chosen an inner route in the first place and avoided a second public inquiry. However, the inspector at the first inquiry considered the alternative routes to be "nicely balanced". Perhaps the western section of Ipswich by-pass should rather be seen as a good example of our public inquiry system with ministers being prepared to change (and delay) their proposals in response to objections.

30. That such delay (to the western section) did not affect the progress of the critical southern section, also demonstrates the wisdom of splitting a by-pass system into discrete lengths for the statutory procedures.

ACKNOWLEDGEMENT

Permission to publish this paper was kindly granted by Mr. D.I. Evans, Director Eastern Regional Office (Transport).

REFERENCES
1. White Paper "Industrial and Regional Development" H.M.S.O., 1972, para. 56.
2. White Paper "Policy for Roads: England 1980" Cmnd. 7908, H.M.S.O., June 1980, para. 4.
3. Technical Memorandum H6/74 "Design flows for motorways and rural all-purpose roads", Department of the Environment, August 1974.

Discussion on Papers 7–9

INTRODUCTION TO PAPER 9 BY L. M. ELLISS
It is arguable whether the western section of the Ipswich bypass should be classified as a major missing link. Although it may not perhaps be regarded as 'major' in the national context, its absence was highly significant for the residents of Bramford and Sproughton and also for drivers (particularly of heavy goods vehicles) travelling between the Midlands, the North and the Haven ports.

It is not any longer 'missing', however, having been opened some four months early in October 1985, but a case study of the history of the scheme provides a good example of the reasons for some of the other gaps in our primary route network, in spite of apparent government determination to complete them.

Although too late to have affected this particular scheme, the Review of Highway Inquiry procedures (ref. 4) did recognise the considerable advantage of holding an inquiry concurrently into more than one route, but confirmed that the Secretary of State would continue to publish orders for only one preferred scheme. It may be worthwhile to consider whether the existing legislative framework should be changed to allow either of two alternative routes of near equal merit to be confirmed following a public inquiry into both of them concurrently.

The contractor for Ipswich bypass, western section (Balfour Beatty Construction Ltd) is to be commended for completing the works some four months early and particularly for producing a concrete carriageway with the best riding quality hitherto recorded in the United Kingdom, despite using the Department of Transport's new C40 specification which has given difficulties elsewhere. An average Q value of 11 was registered by the Cement & Concrete Association's multiwheel profilometer – the range 50–75 normally being regarded as 'good' and 25 being classed as 'very good'.

One structure which may have contributed to the contractor's ability to finish the contract early is the railway arch at Bramford, which received the Institution's East Anglian Association 1985 Merit Award. At Bramford, where the bypass crosses the London–Norwich railway line at a height of some

11 m, a 15 m span, two-pinned arch was cast against a permanent corrugated steel liner before being backfilled with pulverised fuel ash. The 63 m long liner was rapidly erected during night possession of the tracks and provided an effective safety barrier for subsequent construction activities.

Reference
Ref. 4. "Report on the Review of Highway Inquiry Procedures". Cmnd 7133. HMSO, London, 1978, April, Para. 29.

MR V. E. JONES, Hereford and Worcester County Council
With reference to Paper 7, I would endorse Mr Turner's remarks in paragraph 14 about the need to identify some reserve capacity in the primary route network. He refers to an example on the A38 (part trunk, part county principal road within Worcestershire) which runs parallel to and just to the east of the M5 north of Bromsgrove. The A38 forms a principal diversion route for light vehicles from the M5 which is currently being reconstructed and widened to dual three lanes north of Junction 5.

The environmental and traffic effects of the diverted traffic had been anticipated by a rigorous traffic management exercise undertaken by the County Council in close collaboration with the West Midlands Regional Office. The works included the introduction of traffic signals to enable side-road traffic to emerge safely and to assist pedestrians to cross the road with a degree of protection. A principal objective had been to optimise the traffic capacity of the diversion route by imposition of a turning ban and the restriction on parking.

To summarise, I would maintain that major maintenance operations on motorways and heavily trafficked links of the primary route network call for the identification of suitable diversion routes and the use of traffic management techniques to optimise their capacity. Such measures need to be anticipated well in advance if they are to be introduced before diversion of traffic occurs.

DR S. T. ATKINS, Greater London Council
Under-valuation of benefits from road construction
I firmly agree with the statement in Mr Turner's synopsis that 'evaluating the real benefits of roads and facilities (to users) is a difficult business which has not yet been mastered', although I take some exception to the phrase 'to users' as I believe that public investment should be assessed by reference to the impacts on society as a whole and not just on the users.

However, the various claims, assertions and suggestions of additional economic benefits, made by various speakers during previous sessions, and again both in Mr Turner's Paper and in

Mr Dawswell's comment of 'colossal benefits', do require much closer examination. I believe that some of the more cautionary comments from the consultants, Mr G. Jones (Wootton Jeffreys) and Dr D. Coombe (Halcrow Fox), in an earlier discussion, should be noted before jumping to conclusions that the economic benefits of road construction are being significantly underestimated.

The various statements in support of 'extra benefits can be categorised under four headings, namely

(1) restructuring of industry to reduce performance costs
(2) reliability of travel times
(3) 'additionality' of benefits from network completion
(4) developmental benefits - Mr Turner's 'glow round the wire'.

There is insufficient time to examine all of these here, and my comments on the difficulties of evaluating these suggested benefits will focus on item (1), the restructuring effects.

Direct user benefits do act as a proxy for all subsequent effects that take place through the market mechanism. It is only those outside of this process that might properly be called additional. Prudent restructuring of industry involves taking advantage of other production and technological advances as well. Dr Coombe has already commented that there is a very great danger of double-counting if all restructuring benefits are attributed to road construction. There is also considerable potential for 'creative accounting ' by considering the effects 'down-the-line' rather than in their initial impact.

There are discontinuities in freight operations. However, it is the statistical distribution of those discontinuities which is relevant, as only a proportion of them will be released by any particular scheme (for example, some time savings cannot be redeployed effectively)

As the trunk road network nears completion, the principal restructuring opportunities are likely to have taken place. Subsequent schemes are likely to have less influence and affect fewer firms.

Not all industries are predisposed to restructuring, notably those with high site-specific investment. Only firms with a relatively high proportion of transport costs (for example, food retailing) are suited to such action; and as transport costs represent, on average, just 5 per cent of industrial costs, the opportunities may be restricted.

Restructuring often trades reduced production costs for increased transport costs. Mr Turner's statistics refer to tonne-km, but examination of tonnes lifted reveals that less freight is being moved, but over greater distances. Usually, such restructuring involves job losses. Public investment appraisal must consider the costs and benefits to society as a whole, not just to companies and individuals. Additional unemployment places both financial and social costs on society. It may be that these additional costs are reflected in the higher taxation necessary to support these unfortunate people, thus counteracting the benefits from reduced transport costs.

Restructuring frequently involves relocation of employment to suburban fringes and green-field sites, where other factors, such as cheaper land, labour and development costs, are more influential. It is feared that in London this may occur to the detriment of the inner city where significant national and local authority resources are being invested to alleviate unemployment and poor social conditions. Is such relocation truly an economic benefit in the widest sense?

Finally, even if freight benefits were increased by the full 30-50 per cent suggested by Dr Quarmby, this may not have a dramatic effect on the NPV of schemes, given the relatively low proportion of traffic flows represented.

This is not to deny that such restructuring benefits might exist in certain cases, but merely to point out that the issues are of considerable complexity. I believe the same applies to the other items (2)-(4) given previously.

As someone from an academic background, I cannot avoid the seemingly inevitable academic conclusion that further research is necessary. However, I would like to make a plea that industry should consider the evidence from such research with an open mind and should not prejudge the results as Mr Turner (and others) would appear to have done.

Comprehensiveness of transport planning

The Greater London Council is not against road construction. Indeed, the Rochester Way Relief Road, to which reference has twice been made, is a GLC scheme. It is also interesting to note, in the context of criticisms of 'computer-based' forecasting methods, that the predictions for this relief road involved manual assignment, a sensible human intervention in an otherwise mechanistic process.

Mr Dawswell's Paper illustrates failure to consider transport provision comprehensively. This is what this conference is about - a network review. However, in London it is insufficient to consider roads in isolation. A key word in transport planning is comprehensive. Transport planning should be comprehensive in terms of the interactions between land use and transport; comprehensive in terms of relevant spatral coverage and comprehensive across modes. Freight shipments share the roads with private cars and the level of service provided by public transport affects car traffic volumes. It can be shown, both theoretically and, I believe, in practical terms, that the best way to speed up traffic in London is through improvements to the public transport services. The removal of subsidy distortions (which mean that in excess of 75 per cent of cars entering London in the morning peak are in receipt of company assistance with motoring costs) could also be highly beneficial.

Do we see any evidence of a comprehensive approach to

transport by the Government? No - quite the opposite. The present Government has caused the separation in London of public transport policy from roads policy and its removal from democratic control. It is shortly to abolish the strategic authority responsible for the integration of land use and transport policy. It will (except in London) negate the benefits of integrated passenger transport networks through deregulation. These moves are hardly conducive to a comprehensive view of transport planning.

MR G. R. CAMERON, Warwickshire County Council, Department of Planning and Transportation
On new single carriageways in Warwickshire, lorry speeds will be somewhat higher than 25 mph, on account of the fact that communities will be bypassed by these new roads. Dualling can be justified only for a limited part of the County's primary route network.

It is possible for objectors to achieve a reappraisal of the design, and this is no bad thing. Counties and Boroughs should have a view on the alternatives, and on any other views, put forward by non-statutory objectors.

MR T. PUMFFREY, Department of Transport
As I am associated with three of the projects mentioned in Paper 8, I feel some further elucidation is necessary.

The schemes to which reference is made are predominantly urban, or of an urban nature; and the particular point Mr Dawswell makes on statutory procedures does need answering. Firstly, though, it must be stressed that urban schemes are, by their very nature, much more complex than inter-urban schemes, and the problems are compounded. Both severance and the effect on property are greater, and more people are affected. Furthermore, in London, special categories of land seem always to be involved.

Mr Dawswell decries the accrimony associated with many urban scheme public enquiries. However, there was no accrimony of any kind at the public enquiries into the two major schemes he mentions in his Paper. Mr Dawswell may wish to comment on the early use of Working Parties by the DTp, which may have influenced the outcome of the enquiries. The DTp and Working Parties were able fully to assess the facts behind all the suggested alternative solutions before public consultation. The Working Parties were composed of a number of interested groups. I am sure that this contributed towards the reduction in accrimonious debate and, subsequently, in the enquiry time as a whole. In fact, the Greater London Council was involved with these Working Parties and supported the schemes, in principle.

In conclusion, I wish to raise the question of what powers are necessary to mitigate the impact of urban road schemes. The DTp is at the moment looking into this important aspect.

MAJOR MISSING LINKS

I would ask whether it is felt that our present powers are sufficient and also what additional powers we should have to reduce further the effect of urban schemes. Any comments would be of value.

MR TURNER
Industry has no wish to prejudge any real assessment of benefits. The Freight Transport Association, representing the transport interests of industry, wishes only to see a proper evaluation of industrial road benefits so that the Department of Transport, and ultimately the Treasury, can use this to fight for a fair share of the national cake for road development and maintenance. We sincerely believe that the COBA system does little more than rank schemes competing for the same money, it makes no absolute evaluation of the economic benefit.

Transport and distribution represent 18 per cent of industrial costs and it is important to ensure proper decisions both by industry and government. It is particularly frustrating for industry, however, that while it can decide to buy production equipment or lorries, it cannot decide to buy better roads, no matter how good the economic case. Furthermore, although more than enough taxation is paid in fuel and vehicle duty, it is only government that decides whether that money is spent. That is why we must ensure that those who take these decisions, notably the Treasury, have all the facts to help them.

MR DAWSWELL
With reference to Mr Pumffrey's comments, I must accept that the DTp has done everything possible to speed up the progress of the two schemes. However, it has been able to do this only within the statutory constraints imposed upon it. Although many of the alternative solutions were fully assessed before the Inquiry into the Hackney M11 scheme, this did not prevent organisations such as the GLC from offering further new alternative solutions at the Inquiry itself. Quite clearly this prevented the full consideration which would have been possible during the consultations before the Public Inquiry. My point was not to criticise the DTp's effort in pushing the schemes forward, but rather to criticise the statutory constraints which, in my view, are stifling the schemes' progress. Mr Pumffrey's final point is of course valid - if powers to mitigate problems were greater, it would be much easier to resolve individual objections and thus to speed up procedures.

Dr Atkins's criticisms of my comments in relation to the benefits are made in a context which does not take account of the road network which has already been constructed to date. It is true that I am able to refer to a missing link only because roads at either end of that link are in use by the public. In road planning, one must surely consider what should be done only in the light of what has already been done and cannot be undone.

MR A. WHITFIELD, Department of the Environment and Transport - West Midlands Region
With reference to Paper 9, I would not like Mr Elliss to make too much of routes of equal merit. Given that, at present, it is not possible statutorily to publish two routes, we have tried as best we can to proceed in the way he mentions; just once. The results have not been encouraging. Objectors seized on this opportunity to produce even more options and, finally, the inspector came down to a 'third' alternative in his findings.

My own view is that time should be spent early on carrying out a very careful examination of all the possibilities, whereupon a choice should be made which should be adhered to. In pursuing that choice, I see nothing wrong in making the arguments known and in encouraging public support wherever it can be found.

I would suggest that all effort expended in the attempt to capture the minds, and the attendance at public meetings, of the millions of road users who might just lend support to our proposals for new routes, is time well spent.

10. Assessment of urban roads

T. E. H. WILLIAMS, CBE, MSc, PhD, FICE, FIHT, FCIT, Research Professor, Department of Civil Engineering, University of Southampton

SYNOPSIS. This paper is concerned with the formulation of assessment procedures which should be associated with development proposals for main roads in major urban areas. It considers the inter-relationships between the functional and socio-economic issues which arise as a result of complex traffic requirements.

A major objective of assessment computations for urban roads and their traffic is to present, in an intelligible Framework format, the collated items which relate to strategic policies in land use development and transportation systems, as well as detailed benefits and costs of alternative development schemes. Townscape issues should be accommodated, to ensure a balance between function and aesthetics.

The end-products of professional assessments should continue to form a primary basis for the national allocation of financial resources. They should also provide the guidelines for local solutions, with particular reference to value-for-money. Economic assessments, including the traffic and construction elements, should retain their importance in those procedures, provided they are used within the context of a comprehensive Framework.

Public consultations, through open community meetings, are of paramount importance. Their effectiveness is more likely to be improved if the discussions are related to the results of comprehensible assessments. The opinions of the motoring public, freight transport organisations and public transport operators should be obtained, and given due weight, together with those offered by other organisations: that should apply during each stage of planning and design. Initial exploratory consultations should require the availability of less rigorous information than later meetings.

SCENARIO

1. The problems associated with investment programmes in main roads in our conurbations and major towns present complex issues, which range from strategic policies on land use development and transportation to construction details

and traffic control. The inter-relationships of the characteristics of urban land use development and traffic are interlocked by the intricacies of the patterns of human activities, within the lattice of roads and railways which form such a vital part of our national heritage and economy.

2. In recent decades, the awareness and involvement of people, other than politicians and members of the professions responsible for infrastructure development in urban areas, in matters involving major roads proposals, reflect their concern for their ways of life and the character of their surroundings. The public consultation meetings and Public Inquiries often highlight the sources of conflicting debates on mobility and environment.

3. The crux of the argument so often heard in favour of nil or minimum development, in terms of construction of urban main roads, is that route improvements which produce more traffic capacity will result in more private traffic, which thus defeats the purposes of the investment. To ignore that fact would be irresponsible and myopic, whilst one should also recognise the persistence demonstrated daily by drivers of cars, commercial vehicles, taxis, cyclists and pedestrians in their choices of routes along and across urban main roads.

4. Those phenomena of urban life clearly require the optimisation of traffic on existing main roads, with various blends of construction, traffic engineering and control. Allied with that requirement is the critical importance of an awareness of community and architectural factors, which should be accommodated in phased development projects. The formulation of assessment procedures should reflect those facts and incorporate quantitative and qualitative evaluations to form a basis for comparability of alternative proposals. That aspect is particularly relevant to assessment exercises which also involve investment proposals for public transport.

5. In comparison with inter-urban roads, where the needs of "through" traffic usually predominate over those of all other road users, the greater complexity of problems within the heartlands of major urban areas include those on main roads, other than motorways, where local and non-local vehicles compete for space. The by-products of noise, vibration and atmospheric pollution are obviously undesirable attributes, which are accentuated by the discontinuities of flows on roads which also serve, in some areas, as shopping streets and access routes to adjacent residential properties.

6. The channelling of non-local traffic on main roads continue to present design, safety and environmental challenges which require cooperative inter-professional expertise in town planning, architecture, civil and traffic engineering. The succesful integration of main traffic

routes with buildings and various facilities for pedestrians should be a primary objective for those professions.

7. It is appropriate that the recently celebrated century of motoring should be followed by this Seminar, which considers the vitally important roles of the national main roads and their traffic in the economic and social life of this country. The truly urban scenes present dauntingly difficult boundary conditions for those concerned with the maintenance and improvement of main roads. Because space is at a premium, opportunities for comprehensive treatment of townscapes are usually severely limited, without the surgery of demolition and subsequent reconstruction.

8. With urban motorways, segregation of vehicles from pedestrians and the physical separation of conflicting streams of vehicles avoid the hazards of unexpected situations, due to changes in traffic or design conditions; the element of surprise for the driver is reduced to its practical minimum. In contrast, inconsistencies are associated with all-purpose urban roads, which span in spectrum from the classified "trunk" type to secondary shopping streets and residential access routes; they represent the compromise in standards of design and traffic management which are adopted to meet a variety of needs. There are safety risks associated with the progressive and radical changes, from absolute priority for vehicular movements on urban motorways to mixtures of road space, signal timings and other facilities for vehicular and pedestrian traffic; assessment computations and ratings should encompass safety issues.

9. For the purposes of this paper, I shall confine my thoughts to "main" traffic routes in urban areas, irrespective of their administrative classification in national and local terms. The reason for this decision is primarily due to the blends of "through" and "local" traffic which most main roads carry in our conurbations and major towns. I acknowledge the difficulties associated with the definitions of those two types of traffic in particular locations within some of our conurbations, but the traffic functions of lengths of main roads which form inherited networks, either whole or partial, are usually obvious.

10. In one respect, it simplifies my task because the corridors associated with main roads relate to specific links and junctions, which together form a relatively discrete network. Within those corridors, there are significant problems due to interface movements by traffic on the main road links and other traffic cross routes and the adjacent ones which lie in the same general direction. Those issues are generally confined, in comparison with those associated with the widespread labyrinth of secondary and local roads.

CONURBATION ISSUES

ASSESSMENT CRITERIA

11. With the varied and dynamic backgrounds of the activities of people, in urban settings, an objective of considerable importance is to obtain the most effective solutions to problems produced by the major traffic flows on main roads. It therefore follows that the allocation of financial resources is critical, in a phased investment programme, which is firmly based on logical targets of completed main road networks and integrated land use developments.

12. The methodologies of assessment should be geared to accommodate relevant component items, to assist in decisions on investment criteria as well as the resolution of conflicting political, public, professional opinions and interests. Because of the variety of factors which affect main roads issues, area-by-area within conurbations, flexibility in assessments and solutions are required within feasible limits.

13. It is also desirable to ensure as much consistency as possible in the techniques which are used in analytical computations, in order that their results should provide a basis for engineering and economic comparability, project-by-project in various areas of conurbations and region-by-region throughout the country.

14. The combinations of consistent cores of quantitative analyses, together with flexible and structured evaluations of qualitative and environmental factors, should aim to provide satisfactory foundations for decisions on investment programmes. Conjointly, policies and their particular objectives for urban main roads, in strategic as well as local terms, should form the basis of statements included with the other components of comprehensive assessments.

15. These criteria, in collated presentations of "benefits" and "costs", could be expanded and varied versions of some of those reflected in the components of the current Framework method used, by the Department of Transport, in the assessment of inter-urban trunk roads. Key differences between assessments of urban and inter-urban main roads should include the emphases on policy-related transportation priorities, where complex movements of people and goods have to be accommodated within areas of various densities and architectural character.

16. The computational techniques in traffic prognoses and economic evaluations may have some common denominators, in urban and rural areas, but there should obviously be differences in details, because of the variations in the complexities of their particular problems.

17. In general terms, the criteria associated with the assessments of urban roads proposals may be grouped as follows : -

    (i) <u>Policies</u> : Land Use Development - Transportation Systems - Town Planning - Townscape Architecture - Finance

(ii) <u>Engineering</u> : Networks – Routes – Traffic –
Improvements Feasibility – Construction
– Traffic Engineering: management,
operations, control, access, circulation,
parking – Safety – Design – Costs:
capital, maintenance, operations –
Phased Alternatives: investment
programmes

(iii) <u>Economics</u> : Evaluations – Benefits: road users,
property, development areas – Costs:
engineering, roads users, properties,
tenants' disbenefits – Rates of Financial
Returns

(iv) <u>Socio-Environment</u> : Benefits; integrated planning,
architecture, engineering – Conservation:
communities, buildings, amenities –
Disbenefits: noise, air pollution,
vibration, property, safety, visual
intrusion

18. In each of these groups, there would be involved a series of objectives. They would range from strategic investment policies, functional efficiency and value-for-money projects to aesthetics, safety and the quality of urban life. The extent to which proposed main roads and traffic engineering developments would satisfy the stated objectives should be unambiguously presented and based on professional assessments.

19. The desired retention of adequate public transport systems in conurbations, together with their levels-of-service and fares policies, present a range of economic and operational problems. In the main roads sector, buses and coaches contribute their claims for space and facilities, in competition with commercial vehicles and cars. The results of computations of forecast traffic flows include buses: the adequacy of present methods of doing so should be further explored, with particular reference to their use in assessment exercises.

20. Arguments in favour of traffic restraint measures which normally refer to private cars, support the case for improved public transport systems. In reality, there is a degree of restraint built into any urban main road project these days, because of the limited flexibility of corridor space which is available for major reconstruction works. That inevitably leads to the inclusions in proposals, at the design stages, of essential traffic management techniques and boundary conditions for regulations, for the operational control of traffic streams and the rationing of space for stationary vehicles.

21. A significant characteristic of contemporary traffic engineering in major urban areas is the increasing application of systems of coordinated traffic signals, for the purposes of optimum use of junctions and their linked

roads. The continual expansion of the applications of such systems, to area-wide main road networks, is a realistic prospect. The availability of discrete main road networks would greatly enhance the scope and efficiency of those computer-based techniques.

22. The stimulating prospect of future communications systems which will probably include in-vehicle devices for the purposes of receiving information on traffic congestion, accidents and guidance on alternative routes, is also a factor in favour of established networks of main roads. In particular circumstances, assessment procedures should reflect those prospective developments.

23. The spectrum of issues and problems which I have briefly described contain the ingredients for complexity in assessment methods. The risk of 'over-kill', by excessive statistical details, should be reduced by matching the numbers of items and depths of analyses with the amounts of design details and scale of improvement proposals, at the various stages of preparations of plans.

24. If effective communications are to be realised between the professional teams and the general road users, in discussions of the merits of alternative solutions to road improvement problems, early outline plans should be accompanied by general assessment information. Later stages in the public consultation programme should involve more detailed analyses and results.

ENGINEERING

25. The components of engineering assessments are composed of basic traffic engineering analyses, together with design, construction and other related civil engineering factors. The former span the spectrum of analytical requirements from those associated with traffic and transportation planning alternatives to the technological systems of management and control of traffic flows. The latter contribute the boundary conditions of feasible alternatives in physical developments and their financial implications.

26. Analyses of the causes and effects of present and future traffic flows on main roads are based on the methodologies of quantitative procedures. The use of mathematical models of road traffic have, as a major objective, the testing of alternative proposals for the development of parts of the main roads network. The results obtained from the computations are applied in design during the early stages and later detailed work; cost-benefit analyses and the preparation of phased programmes of implementation of plans and investment of resources.

27. Analytical models of road traffic and transportation systems appear in various guises, most of which are variations on the fundamental theme of a three-phased process concerned with :

Generation (of traffic in relation to land use);

Distribution (of traffic, between centres of activities)
Assignment (of traffic, to specific routes and by various modes of transport)

28. There are intermediate stages of considerable importance, interspersed within this general framework for mathematical models. They provide a fertile field for pragmatic practitioners as well as theoretical research workers. The elegance of some theoretical treatments of these problems occasionally tend to be out of balance with the quality of the raw data which is available for input. There is a continual need for improved matching of the field data with theoretical methods and the progressively finer limits which are available from computers and their related software programmes. The interpretation and use of the information embodied in voluminous matrices of data offer the prospects of statistical indigestion; a significant challenge is the formulation of methods for the presentation of facts and figures in digestible quantities, without camouflage.

29. The applications of the results of traffic computations to networks of main roads, in whole or in part, soon highlight the critical importance of the inter-relationships of Land Use Development Plans and Transportation Systems. In area-wide conurbation terms, a strategic planning objective is usually a relatively coarse main road network, of adequate standards and capacities, together with their feeder routes, which serve to channel and distribute the traffic to other all-purpose roads. Continuity of route and consistency in types of design and techniques of traffic management and control, which underlie the basic engineering approaches to improvement proposals for main roads, are frequently and severely constrained by the character and densities of land use developments of the areas through which they pass, and also serve.

30. The assessment of the engineering feasibility of main roads projects include the effects of the degree of compromise in design and construction which would be a practical necessity, as opposed to desirability, for the purposes of implementation of strategic plans. The greater the compromise, in terms of geometric design, through lower numbers of traffic lanes, less grade-separated interchanges, more at-grade junctions, the greater the requirement for traffic engineering techniques and police enforcement - traffic lane controls, signs, signals, markings, kerbside facilities and parking controls.

31. The amounts of restraint on potential local road users, as a result of compromises in the design of main roads, is probably minimal in many locations; their problems are more likely to be acute on other all-purpose roads in conurbations and major towns. The extra local traffic which is likely to be generated and attracted to improved main roads will reflect the standard of construction of the completed works

and the level of efficiency of the traffic control systems.

32. The types of "generated" traffic would include that which is extra private car journeys during peak periods with the commuter change from public transport. Also some latent extra journeys, during off-peak periods, will probably be realised, to the satisfaction of motorists who wish to be involved in the wide range of human activities which are associated with major urban areas. The particular problems of peak periods of traffic flows, and their management, continually offer scope for research, development and application of traffic engineering techniques.

33. Construction procedures inevitably involve physical upheaval and disruption of normal life, in areas where works occur; main roads offer no exception to that situation. The replacement of cables and pipes associated with services below ground level as well as the protection of adjacent buildings and the rerouting of existing vehicular and pedestrian traffic all contribute to the costs of projects. The phasing of execution of improvement plans, with the incremental investment over many years which is necessary with main roads, add to the engineering problems and their reflection in assessments and evaluations.

34. Existing lengths of main roads which form the links in a network, between interchanges and at-grade junctions, lie within established corridors of travel. They form the spines of the traffic corridors, within which improvement proposals usually are located. Alternatives in planning-design-construction-operation, within the corridor areas, form realistic approaches to development plans.

35. At-grade junctions on main roads networks present sources of capacity restraint for traffic. Their importance is reflected in the research and development effort which is being devoted to the improvement of their geometric design and signal-controlled traffic operations. The work of the Transport and Road Research Laboratory, together with University groups, practitioners, individual research workers and consulting engineers have resulted in highly significant improvements in that field. The perennial problems of junction traffic capacities, in relation to link roads and spaces for vehicles in queues behind stop lines, continue to highlight the importance of their work. The measure of their success are the delays to vehicular traffic, as well as the magnitudes of flows, together with safety of pedestrians, drivers and passengers.

36. Critical and outstanding problems associated with subject areas in the traffic engineering field include : -
(i) Forecasting of traffic growth, with particular reference to urban land use changes
(ii) The mathematical modelling of the different types of main road traffic under congested conditions, as well as accommodating "modal split" issues and variable trip matrices

(iii) The attributes of industrial, commercial and public transport traffic flows on main roads, and their accommodation
(iv) The extension of Urban Traffic Control technology to include Route Guidance systems and incorporate comprehensive pedestrian facilities, wherever possible
(v) The extent of redistribution of traffic flows in areas within and adjacent to well established main roads, as a result of improvements in their layout and traffic control, on links and at junctions.

ECONOMICS

37. The economic component of Assessment is a vital part of the appraisal procedures for main roads. The computations of benefits and costs, in financial terms, serve to indicate the probable rates of economic returns on the significant sums of money involved and thus provide a basis for claims for the investment of public resources.

38. In engineering terms, the availability of cost-benefit evaluations can provide a valuable means of confirming, or otherwise, the advantages of a particular designed proposal, when compared with others. Where alternative improvement proposals for a main road satisfy policy, engineering and environmental criteria, value-for-money remains as a critical issue. To obtain a solution which is functionally satisfactory and at the lowest cost, in terms of real resources to the nation, is obviously of paramount importance. That objective chimes with the traditional civil engineering approach to design problems.

39. Benefits to main road users, from proposed improvements to routes, are normally estimated in terms of savings in journey times, vehicle operating costs and accidents. The computed products use vehicular traffic flows by vehicle type; the reductions in travel times and delays; the values of time for vehicle occupants; the rates of accidents involving personal injuries, based on the recorded police records, for various types of road.

40. Where urban locations of particular types are concerned, pedestrian flows may be significant. Delays experienced by them should, wherever possible, provide an additional item for computation and evaluation. Relevant results from research on delay to pedestrians using at-grade crossings on town traffic routes are already available as useful indicators of the orders of magnitude of the quantities involved.

41. In conurbations and larger towns, industrial developments and general urban renewal programmes will usually benefit from improved main roads, in their immediate areas and their links to the national trunk road network. The evaluation of those benefits, for their use in detailed economic appraisal of proposed projects would require 'before-

and-after' survey results of similar land use changes and developments. There appears to be a shortage of data banks for that purpose.

42. Costs associated with urban main roads improvements include those for engineering design, property and land, construction, maintenance and traffic operations. Disbenefits, in financial terms, occur to some owners or tenants of roadside properties, whereas benefits are experienced by others, in similar locations; this is a difficult and sensitive subject area, involving home owners and tenants in various types of buildings of different ages. Commercial and industrial premises may be less susceptible to loss of benefits, provided adequate access is safeguarded to their properties and parking areas. The additional costs of sound-proofing and noise barriers have to be met, in some locations.

43. In most of these circumstances and procedures, a common denominator of crucial importance is that of potential traffic flows on the affected main roads. Statistical Confidence Limits provide measures of the scatter and uncertainty which may apply to the computed traffic; they should be borne in mind when the results of detailed economic evaluations are assessed. The values of such calculations, in relative terms, as indicators of rates of economic returns and when used as design aids, are obvious. Their consideration, in absolute terms, should be qualified by the attached variations of current field and test data, together with the uncertainties associated with forecast elements of the economic equations. The appropriate values of time for travellers continue to provide difficulties in economic assessments, in attempts at absolute evaluations.

44. Whilst recognising the problems arising from the scatter of real-life data, and their knock-on effects in many sectors of economic assessments, the need for consistency in those particular procedures is essential. Without a consistent methodology, the comparison of the economic attributes of alternative designs would not be possible; that would particularly apply to projects prepared in different regional areas, which have to compete for national finance.

45. In contrast with the case for consistency in economic computations, greater degrees of flexibility may be desirable in the traffic estimation phases. They would encourage continual research and allow the use of improved mathematical model techniques for the production of numbers for design flows of main roads. The computed flows, for off-peak and peak periods, together with the appropriate flow-speed-junction delays relationships should provide the common base for traffic and economic evaluations of alternative designs.

46. In the light of the fact that urban main roads traffic

is influenced by a wide variety of factors, including local rates of growth, the horizon used for economic assessment beyond the year of probable completion and opening of schemes, should reflect the uncertainties produced by those factors. The current "COBA" period of 30 years which is used by the Department of Transport is excessive; 15 years, plus a First Year Rate of Return are more appropriate.

47. The well founded criticisms, aired in many a forum, of the present format of COBA as a suitable method of economic assessment of urban main roads, have common themes of lack of flexibility and true reflections of the characteristics of conurbation and major town traffic. There are opportunities for research and development, to urgently improve that and other possible alternative computer-based cost-benefit programs, to obtain a more satisfactory balance between consistency in technique and flexibility in application.

SOCIO-ENVIRONMENT

48. The effects of main road traffic on the environment of people who live, work and visit urban areas and their facilities provide an interesting range of problems. Most are recognised, and considered nowadays, at all stages of development of main roads improvements, from embryonic plan exercises, through detailed designs and to the implementation proposals of traffic management programmes.

49. The channelling of major traffic flows in main roads inevitably concentrates disbenefits, in environmental terms, along their corridors. Allied with those negative characteristics are the positive benefits experienced by residential, local shopping, cultural and other areas from which extraneous vehicles have been drawn to the main routes. The extent of those traffic catchment areas are often difficult to delineate, although they could prove to be significant in the qualitative sectors of overall assessment. Environmental gains are usually more diffuse and widespread than the losses, from main road improvements.

50. Urban renewal projects, on a comprehensive scale, may offer the opportunities for improved townscape design and construction. The successful integration of improved main roads with adjacent buildings, structures and centres of recreational activities require inter-professional cooperation and imaginative expertise. Aesthetics form a precious aspect of our urban heritage, and add to the stability and quality of life in situations which are frequently hyper-active in other ways.

51. The protection and continuity of activities of urban communities present investment claims of a challenging variety. The barriers formed by main roads and the safety of pedestrians, in their vicinities, provide permutations of design and police problems.

52. Noise, vibration and atmospheric pollution are

quantifiable by-products of trunk road traffic. Together, their values could be used as indices of local discomfort in relation to guidelines on their tolerable and desirable levels. The ranges of frequencies of measurable harmful effects of noise and vibration are of particular importance. The future prospect of reduced air pollution from vehicle exhaust fumes as a result of 'lean-burn' engines and catalytic innovations should improve that situation. In the meantime, noise levels may be an acceptable single index proxy for this group of pollutants.

PRESENTATION

53. The data and information on socio-environmental facts and policy issues, together with the results of engineering and economic analyses should be collated in the form of a Framework. The basic composition of Frameworks should be specifically urban and possibly be expanded versions of the Department of Transport type currently in use for inter-urban trunk roads assessments.

54. Corporate professional use of Frameworks should involve increasing degrees of sophistication of Assessments, with the greater complexity of problems. Initial planning and design exercises may only require macro levels of key information, followed by micro data for detailed design. The involvement of townscape planners and architects should be ensured during each stage of the design programme.

55. The frequency of use of Frameworks information in public consultation exercises should at least coincide with the various stages of planning and design. At an early stage, when macro information is available on the complexity and scale of the proposed main road improvements, open meetings should enable the discussion of general characteristics and boundary conditions involved in the professional efforts. A particular feature of early consultations with the general public should be the explanation of transportation policy and design objectives, with alternative degrees of emphases in main roads proposals on new construction, Traffic Engineering techniques, traffic management regulations and enforcement. I recognise that this procedure is already adopted in inter-urban exercises and many town schemes.

56. The explanations of the issues involved in planning, engineering design and economic resources should also be directed towards motoring organisations, freight transport associations and public transport operators. They should be encouraged to offer constructive comments and participate in public debates, in order that the views of motorised road users should be effectively heard, in addition to those of other organisations. To be satisfactory, public participation in the assessment of main roads plans should involve as comprehensive a sample of road users as possible: the Frameworks should provide the means of structuring

debates in productive directions.

57. When additional work is completed on feasible alternatives for the improvements of main roads, a second opportunity of testing public reactions should be available, by means of another round of open meetings and written submissions. The solution preferred by the professional teams should be presented and illustrated in appropriate detail in architectural-townscape terms as well as engineering facts, figures and layout drawings. The "benefits" as well as "costs" should be clearly stated by means of concise summaries of Framework information. It is hoped that the effectiveness of this public consultation phase would be reflected in less confrontation issues in the formal Public Inquiry proceedings, in due course.

ACKNOWLEDGMENT. This paper is a personal effort but I should acknowledge the possible reflection of some of the opinions held by individual members of the Standing Advisory Committee on Trunk Roads, Department of Transport. I hope that I have incorporated their views in correct context.

RELEVANT PUBLICATIONS
1. DEPARTMENT OF TRANSPORT. Report of Advisory Committee on Trunk Road Assessment. H.M.S.O. 1977
2. DEPARTMENT OF TRANSPORT. Traffic Appraisal Manual. APM Div. 2, Marsham St., London, SW1P 3EB
3. DEPARTMENT OF TRANSPORT. COBA 9 MANUAL. APM
4. DEPARTMENT OF TRANSPORT. Manual of Environmental Appraisal. APM
5. DEPARTMENT OF TRANSPORT. Frameworks for Trunk Road Appraisal. Dept. Standard TD/12/83. APM. 1983
6. COOMBE, D.R.; DALY, P.N.; WHITMARSH, R.J.; EVANS, R.C.; PALING, R.S.; FINGLAND, D.; GURNEY, A. Urban Road Appraisal Proc. PTRC Conference. 1985.

11. Primary routes in urban areas

S. N. MUSTOW, BSc, FEng, FICE, FIHT, County Surveyor, West Midlands County Council

SYNOPSIS. A paper directed to examining the role and importance of the Primary Route Network (PRN) in the major conurbations in the light of their movement, characteristics and demography. Developments in route classification are examined and some key issues that have to be addressed in the ongoing development of the PRN are considered.
Attention is drawn to the importance of increasing transport investment in the conurbations and a case is argued for expenditure on the PRN to be matched by investment in other urban roads or similar importance.

INTRODUCTION
1. The major conurbations provide the focii for much of the traffic using the nation's transport systems, whilst the increasing sophistication and dispersion of industry has made the national road and rail networks essential elements in the manufacturing and distribution processes. The great progress made in the development of the national motorway and trunk road systems has done much to improve inter-urban transport but developments in urban areas have not been as rapid or widespread. There are a variety of reasons for this situation, environmental, political and social in character and this paper examines the nature of urban movement, the importance of road transport and the place of the Primary Route Network (PRN). Specific attention is given to network definition, the need for investment and practical issues associated with the developing Primary Route Network.

Table 1 Area Population and Employment Density 1971 and 1981

	Area (Hectares)	Population Density 1971	Population Density 1981	Employment Density 1971	Employment Density 1981
England & Wales	15,120,700	3.22	3.25	1.35	1.38
Conurbations (Mets & GLC)					
Greater London	157,944	47.18	42.50	21.60	19.52
Greater Man.	128,674	21.21	20.17	20.2	8.47
Merseyside	65,202	25.41	23.19	10.06	9.06
S. Yorkshire	156,049	8.48	8.36	3.62	3.42
W. Yorkshire	203,912	10.14	9.99	4.28	4.18
Tyne & Wear	54,006	22.44	21.16	9.00	8.47
West Midlands	89,943	30.46	29.45	13.76	12.06
Conurbations	855,730	22.14	20.98	9.80	8.99
England & Wales excl Conurbation	14,264,970	2.07	2.19	0.85	0.92

Source : Source 1971 and 1981 Census

PAPER 11: MUSTOW

CONURBATION CHARACTERISTICS
The Urban System

2. The major conurbations where towns, villages and cities coalesce form a densely developed conglomerate of activities. Taken together the six Metropolitan Counties and Greater London account for 40% of the population and employment of England and Wales within 5.5% of its land area (Table 1). Taking the densities of population and employment as proxies for activity, the land area of the remainder of England and Wales would have to be shrunk by a factor of 10 to achieve the same intensity of interaction. It follows that if major cities such as Nottingham, Cardiff or Southampton were transferred to the urban score the factor would be even greater.

3. Conurbation land-uses are not evenly distributed and are such that factories lie cheek by jowl with houses, shops and offices and their varying traffic generations give rise to complex movement patterns which change over time under the influences of economic and social change.

4. Density is therefore not static and over the last quarter of a century slum clearance, town centre redevelopment, industrial decline and increasing mobility have exercised major influences contributing to the gradual decline of conurbation populations, the decentralisation of activities and the further isolation of less mobile lower income groups in the inner areas of the conurbations.

5. Movement in the conurbations is thus the product of a system of great complexity in which the relationships and interactions have to be understood if adequate provision is to be made for the future. It follows that the various transport systems moving people and goods must be understood as interdependent.

Employment

6. A study of landuse in the major conurbations reveals the close relationship of transport systems with the development of industry and housing. Major rivers attracted development as they formed the major transport link for the movement of goods whilst the later building of canals, railways and roads was also reflected in development. Birmingham and the Black Country are examples of the major significance of the canal system (Fig 1). The key factor in industrialists' minds today is access to the national motorway system and its importance is demonstrated by the development that has taken place near junctions. Employment in the six Metropolitan regions has declined by nearly 950,000 or 5.8% since 1971 leaving large areas of land derelict in the urban heartlands and the process of reclamation and redevelopment will be unsuccessful without corresponding improvement in road access. Ironically problems are made more difficult by the development of ring motorways around the conurbations which further tip the balance against the attractiveness of inner area sites for industrial/commercial redevelopment in favour of peripheral sites close to the new roads.

CONURBATION ISSUES

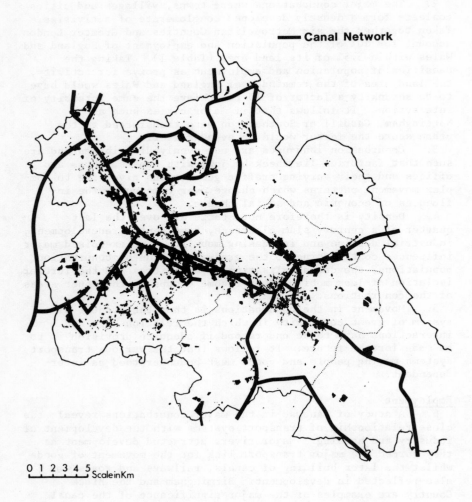

Fig. 1. The distribution of industrial land in the West Midlands Conurbation in 1975

Roads and their Usage

7. The conurbations contain about 16% of the road system of England and Wales with similar percentages of motorways, trunk and principal roads and primary route destinations (Table 2). No consistent specific data is available of traffic in the conurbations but the Department of Transport, Transport Statistics 1974-84*[1] show that some 54% of the 240 billion veh.kms. travelled in England and Wales in 1983 was on built up roads. On average built up roads carry about 2500 vpd compared with 1500 vpd on non-built up roads.

8. A continuing thread in government transport policy in recent years has been the need to provide by-passes to take heavy through traffic away from environmentally sensitive villages, towns, shopping and residential areas whilst improving access for industry and commerce. Heavy goods vehicles are an increasing problem in the major urban areas and rail freight into the conurbations has changed in character and declined over a number of years. There are many reasons for these changes but it would be mistaken to assume that they can be reversed and it would be wrong too to assume that the majority of HGV movements in the conurbations are through traffic. A recent study in Dudley for the proposed extension of the Black Country Route indicated that out of over 5000 HGV's per day using roads near the centre of the town, 86% had an origin or destination within the County boundary and 60% of these had both origin and destination within that boundary. Land-use distribution and freight movements preclude the general application of lorry route or by pass philosophies in the conurbations.

9. Other categories of through traffic constitute a very small proportion of total movement in the conurbations and this is illustrated by the fact that on a typical weekday at the boundary of the West Midlands, only 17% of traffic on main roads is through traffic whilst further into the conurbation this component becomes insignificant.

10. Conurbations provide employment for surrounding rural areas and in consequence commuting is an important element of urban movement. In the West Midlands over 40% of employed residents in surrounding areas commute into the conurbation. Despite increasing unemployment commuting has increased, particularly by car.

11. The density of the urban road network results in a correspondingly high density of junctions which determine the capacity of the system. As traffic approaches capacity, small increases in flow at these points result in large increases in delay. The greater number of conflicting movements at junctions results in higher accident rates than on road links.

12. Car ownership and usage for journey to work is significantly lower within the conurbations than in the rest of England and Wales but nevertheless continues to grow. The lowest levels of ownership are concentrated in the areas of greatest social need and it is these people that are particularly dependent on public transport. (Table 3).

CONURBATION ISSUES

Table 2 Share of the Road System

Area	Total Length of Road		Motorways, Trunk & Principal Roads		PR Destinations	
	kms	%	kms	%	No	%
England & Wales	307831	100	39585	100	351	100
Conurbations	48968	16	6062	15	59	17

Sources : CIPFA Highways and Transportation Estimates

Table 3 Households England and Wales with and without Cars 1971 and 1981

	England and Wales		Met Regions		Other Regions	
	1971	1981	1971	1981	1971	1981
Total Households	16,509,905	17,706,496	12,614,865	13,092,247	3,895,040	4,614,249
Households without Cars	7,952,485	6,824,470	6,256,705	5,243,427	1,695,780	1,581,043
Households with Cars	8,557,420	10,882,022	6,358,160	7,848,820	2,199,260	3,033,206
% with Cars	51.8	61.5	50.4	60.0	56.5	65.7

Source : 1971 and 1981 Census.

Table 4 Extent of Road Classification in 1966

Class	Mileage	Percentage Distribution of Costs	
		Ministry	Authority
Motorway	391	100	-
Trunk	8,349	100	-
Class I	19,860	75	25
Class II	17,642	60	40
Class III	48,998	50	50
Unclassified	105,835	-	100

Public Transport Usage

13. Public transport is vital to the urban system and without it the pressure for road space would be intolerable leading either to an endless pursuit of capacity with all that means in terms of environmental damage or to growing congestion and further migration of business. Most of the conurbations exhibit ridership decline and there is little doubt that major investment will be required in LRT and metro systems if long term improvement is to be achieved whatever the results of the Buses Bill.*²

CONURBATION ROAD NETWORK
Network Definition

14. Ever since the expanding use of the motor vehicle led to the establishment of a Ministry of Transport and the definition of the first national road classification system, networks have been developed for a variety of specific purposes including route signing, grant distribution, network planning, road maintenance, lorry and 'bus routing. Network definition is difficult in urban areas due to the density of the system and has often had little to do with the amounts of traffic using the roads or relationships with other networks. The following paragraphs illustrate some of the resulting problems :-

15. The initial classified network was modified over the years by the creation of Trunk Roads (1936) and Class III Roads (1946) so that by 1966 when the motorway programme was in full swing the general definitions were :-

Motorway/Trunk : routes of national importance.

Class I roads : roads (excluding trunk) connecting large towns, of special importance to through traffic and/or of regional importance.

Class II roads : important connectors between smaller population centres.

Class III roads : roads of more than local significance.

16. For many years this network was of great importance as it formed the basis of the route signing system and of government grant distribution to highway authorities (Table 4). But though the signing system and the grant rules have long since changed these networks still exist in law. Trunk Roads were not generally designated in County Boroughs and this meant that these authorities had little opportunity of obtaining the 100% Trunk Road grant. Furthermore the presence of County Boroughs led to the existence of disconnected sections of Trunk Roads in the conurbations, some of which exist to this day.

17. Principal roads made their appearance in 1967 as part of a modified classification and grant system. They comprised most Class I roads and the classifications became :-

CONURBATION ISSUES

   (i) Trunk Road Network. Strategic routes of national importance catering for the through movement of long distance traffic.
   (ii) Classified Principal Road Network. Second-tier road system incorporating regional and district distributor routes and complementing the trunk road network.
   (iii) Classified Non-Principal Road Network. A subsidiary system of roads of local importance giving access to industrial, commercial and residential sectors.
   (iv) Unclassified Road Network. Feeder roads including minor rural roads as well as urban estate roads.

   N.B. It will be noted that the old classified roads which did not become principal roads were absorbed but not extinguished by this system.

18. The Primary Route system which was introduced in 1964 as a product of the Worboys Report (1963)*[3] recommended a consistent standard of route signing for long distance traffic movements using a limited list of Primary destinations. The network comprised trunk roads and a large proportion of Class I roads but did not affect the status of other classified roads and was not used for grant allocation purposes. There was much to commend the system but conurbation authorities were generally slow to implement the proposals due to the costs involved and the problems of cross boundary co-ordination. Since 1974 the West Midlands County Council has designed and implemented a comprehensive re-signing scheme including Primary Routes and this work is now complete, but there are still parts of the country where the appropriate signing has not been provided.

19. For transport planning purposes "Roads in Urban Areas" (1966)*[4] advocated a road hierarchy based on primary distributors, district distributors, local distributors and access roads with recommendations for control of access onto each type of road. This type of hierarchy had no consistent relationship with the road classifications mentioned earlier or with the required approach for the strategic transportation policy chapters of Structure Plans under which roads are to be categorised in a hierarchical system involving "primary routes".

20. In terms of planning control, local highway authorities have powers to give directions restricting the grant of planning permission for certain specified types of development adjacent to the classified road network and the local planning authority is empowered to restrict development on unclassified roads on highway and traffic grounds, but as explained earlier highway planning is undertaken on the basis of road hier - archies defined on different premises which can lead to misunderstanding and confusion.

21. There is no clear relationship between the volume and type of traffic using urban routes and their classification so that in the West Midlands, for example, 85% of the 380 km network of primary routes carries in excess of 10000 vehicles or

1000 HGV's per day, whilst over 600 km of non-Primary Routes carry similar levels of traffic (Fig 2). Even within the designated networks themselves there is considerable variation of traffic flow. It is in this generally unclear context that the ongoing development of the Primary Route Network needs to be considered.

THE PLACE OF PRIMARY ROUTE NETWORK IN THE NATIONAL ROAD SYSTEM

22. The PRN has been increasingly seen as a focus for long term capital investment as well as a national system of routes for longer distance traffic complementing the Trunk Road system. The process commenced with the Standing Conference on London and South East Regional Planning*[5], continued with the County Surveyors' Society through the Publication of Transport for the 80's*[6] and concluded with government approval and the modification of the grant system to direct finance towards the specific development of the PRN and roads of strategic significance. In reviewing the system the government declared*[7] "our aim in the review has been to bring the network up to date taking fully into account the benefits conferred on the longer distance traveller by recent developments, particularly the growth of the motorway system. The guiding principles have been to :-

(a) retain the PRN as the system of routes serving longer journeys - rather than the needs of local traffic - between places of traffic importance, primary destinations;

(b) adjust the PRN only where there are compelling reasons because changes are confusing to the motorist and expensive to implement".

23. Welcomed by the Association of County Councils, the government's proposals for the development of the Primary Route Network had only conditional approval from the Association of Metropolitan Authorities (representing the great conurbations) and it is important to understand the differing approaches involved when considering the future of the PRN. Both sides were agreed on the necessity of developing the PRN as a through-route signing system and conurbation authorities were well aware of the vital importance of industrial and commercial access to the motorway system but it was on future investment that views differed. From the rural county point of view there was much to be said for the development of a consistent route framework for ongoing investment. It would facilitate early route definition across boundaries and help in the development of regional strategies, structure and local planning and the location of commerce and industry. To the urban authorities however the notion of considering the PRN in isolation from the rest of the urban transport system was a recipe for unbalanced decision making and the syphoning of funds from the deprived conurbations to the rural authorities. This last

CONURBATION ISSUES

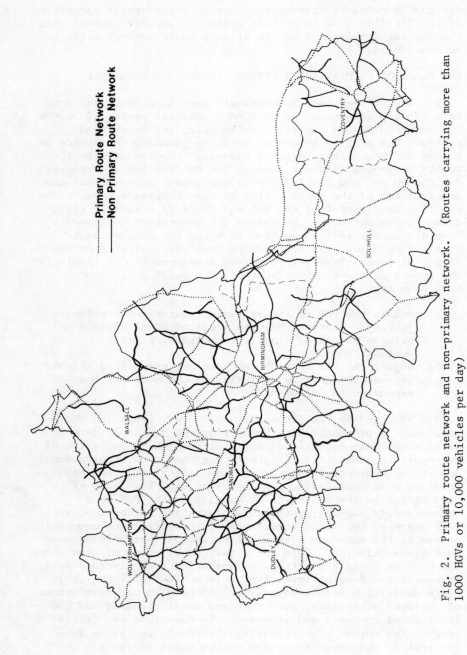

Fig. 2. Primary route network and non-primary network. (Routes carrying more than 1000 HGVs or 10,000 vehicles per day)

point was given added force by the recognition that many unclassified roads in urban areas carry more traffic than Primary Routes in rural areas and would not necessarily rank for grant under the new system.

24. In the conurbation context the following are some of the issues that have to be addressed in planning the ongoing development of the PRN :-

(i) the role and extent of the urban network having regard to the total transport system, road traffic and the need to improve industrial and commercial access.
(ii) the extent of the non-PRN that will rank for grant and the criteria by which it will be judged.
(iii) the basis of programme development and government grant distribution for urban and inter-urban PRN development.
(iv) the means of evaluating and ranking urban PRN proposals.
(v) design standards for the system and the administrative context.
(vi) the need for a comprehensive review of road classification systems in the interests of clarity and efficiency.

The Role and Extent of the Urban PRN

25. In the urban context the PRN fulfils a vital role as the link between the inter-urban system and journey origins and destinations; furthermore the conurbations include a number of primary destinations. It must be remembered however that the PRN is part of the total internal distribution system and as such carries a large volume of traffic that does not cross the conurbation boundaries. It is this duality of function that demands that PRN development is considered in the context of total transport need.

26. As part of the recent government review, plans have been prepared for updating and extending the PRN by representatives of central and local government and an important consideration has been the relative density of the networks in urban and inter-urban situations. "Transport for the 80's" was based on the general thesis that "a national network of roads should be defined that would :-

(a) Provide accessibility to centres of population, industrial generators and outlying rural areas.
(b) Cater for longer distance traffic movements between the areas.

Towns with a population of 25,000 and above would be served by the network being within 5 miles of a primary route and having reasonable access to it --- the network must give entry into larger towns, cities and Metropolitan areas ---".

27. These principles are a useful starting point for defining a Primary Route Network in rural areas but they are not so helpful in the conurbations so it is necessary to keep in mind that it is these conurbations which are the major

generators of traffic and which need improved access if they are to revive. The representatives regarded it as important not to so slavishly apply the Transport for the 80's criteria that the inter-urban PRN density was totally at odds with that agreed for the conurbations. In the event the networks for the six metropolitan counties were based on the contents of their Structure Plans and comprise about 5% of the conurbation road mileage. A key issue outstanding is the PRN for Greater London. In the response of the Association of Metropolitan Authorities to the Government proposals it was stated that "due to the complex nature of the highway network it is inappropriate to attempt to define a primary network for London". As a result an updated version of "Routes for Longer Distance Traffic in South East England"*[5] was used to give an indication of the scale of investment required. There may be political difficulties in the definition of a PRN for Greater London but it is difficult to see how the task can be regarded as "inappropriate" in technical terms.

Routes of Similar Importance to the PRN

28. Evidence has already been advanced to show that a considerable number of non-PRN roads in urban areas carry volumes and proportions of traffic that are similar to Primary Routes and should therefore attract government grant for their improvement. When announcing changes in Transport Supplementary Grant in 1984 the Secretary of State for Transport indicated that grant was to be limited to projects on the Primary Route Network or to major roads of strategic significance which may carry volumes of long distance traffic or by-pass communities to relieve them of heavy through traffic. It is essential that these definitions are generously interpreted if the overall road system for the conurbations is to get the attention its trafficking justifies. If problems develop in this regard the Primary Route Network should be further extended to include other roads of similar importance. It is worth noting here that when Transport Supplementary Grant was instituted in company with the Transport Policies and Programme System in the early 1970's the intention was to give local Highway Authorities maximum flexibility to direct capital to projects of greatest need irrespective of road classification. Whatever the broader justification for the changes in grant arrangements the loss of the flexibility provided by a block grant is a serious matter for local authorities and welcome as specific grant for certain routes maybe it is no substitute for that lost flexibility.

Urban/Inter-Urban Programmes and Grant

29. Transport for the 80's advanced the view that if the PRN (excluding London) is to be brought to the requisite standard by 1992 total transport expenditure must be increased by about £68.5M p.a. (November 1979 prices) or 2.5% p.a. The Association of Metropolitan Authorities has since shown that the cost of improving Primary Routes in the six metropolitan

areas is £528M (Approx. £53M p.a.) and has argued that this
expenditure should be paralleled by major expenditure on public
transport improvements as a total transport package. Road and
public transport improvements in the GLC were tentatively
estimated at £1800M (of which £450M related to roads) over the
period to 1992. These considerations are crucial for the
conurbations as since 1974 there has been expansion of the
motorway and trunk road programmes at the expense of local
transport expenditure to the extent that over the last 5 years
expenditure on national roads has increased by 60% in cash
terms compared with only 20% on local roads. There has also
been evidence of the balance of expenditure within the local
transport sector being tipped in favour of inter-urban
authorities as against urban authorities via the Rate Support
Grant and Transport Supplementary Grant allocations. For
example the accepted expenditure for Transport Supplementary
Grant for 1985/6 was 35% higher per head of population in the
Shire Counties than the Metropolitan Counties. In these circumstances it is particularly important that the criteria
governing grant allocation for the developing PRN programmes
are seen to operate in an even-handed way as between inter-
urban and urban authorities.

Evaluation of Urban PRN Proposals
30. Each of the major conurbations is covered by comprehensive land use - transportation studies incorporating up-
dated models describing the systems and allowing alternative
transportation strategies to be tested. There is thus an
effective basic framework on which to test the PRN and
determine priorities for improvement. However it is essential
that proposals are tested against environmental, developmental
and other criteria as well as the usual traffic and economic
considerations. Most urban authorities have developed weighted
balance sheet type evaluation systems to justify and rank their
projects because the COBA 9*[8] economic evaluation developed by
the Department of Transport for freestanding Trunk Road
projects is not appropriate for the evaluation of schemes in
dense urban systems - a point which is recognised in the COBA 9
instruction manual. Inevitably these approaches involve a
combination of objective and subjective elements and differing
approaches between the conurbations.

31. The need for new ways of evaluating road projects was
well made by the County Surveyors' Society in Transport for
the 80's as follows :- "In recent years the majority of individual highway improvements and new roads have shown economic
returns on capital investment which warranted construction on
this criteria alone. The Society believes that present cost
benefit analysis techniques take reasonable account of time and
accidents saved on a scheme-by-scheme basis and, therefore,
provide an acceptable measure of the value of most highway
projects, examined in isolation.
But these techniques cannot evaluate the additional benefits
which would stem from the existence of a complete national

network of good standard routes. International investment must surely be attracted to a country having an efficient transport infrastructure, if not complete, at least firmly programmed. Industry and commerce would reduce costs by being able to plan and develop their transport operations with more precision and certainty of delivery. Additionally, highway schemes often improve the environment in towns and villages and provide an opportunity to introduce traffic management measures to exclude extraneous traffic from sensitive areas. All these benefits are recognised in principle, though unquantified in economic terms.

In summary, we believe that current methods of analysis consistently underestimate the value of highway investment. It is to be hoped that some system of more general application will result from the work of SACTRA which is more applicable to ongoing work on the PRN for urban areas than the manual of environmental appraisal.

Design Standards
32. The various estimates of cost of updating the PRN have been based on the principle of providing the most appropriate facility for the particular situation leading to the use of single carriageway and dual carriageway (2 x 2 and 2 x 3 lane) solutions and reduced levels of service for certain urban schemes. In this way it is possible to demonstrate the greatest justification for individual schemes and any move away from the principle towards say a standard dual 2 lane carriageway approach would be likely to lead to the diversion of funds from areas of greatest need.

Administrative Context
33. It has been argued in this paper that road developments in major conurbations must not be considered in isolation from the development and use of the rest of the transport system. Whatever the over optimistic expectations that have been expressed about planning over the last two decades the principle has been regarded as self-evident by a range of bodies such as the House of Commons Transport Committee* as well as all the professionals involved. In national transport terms the Department of Transport has a remit for all forms of transport but from April 1986 there will be no local authority with comparable responsibilities for the conurbations. Highway authorities will have to begin to coordinate their activities in areas where administrative boundaries have little significance and for the most part they will have little power to directly influence public transport policy. Bearing in mind that the PRN is by definition a national system, effective coordination and direction will be essential in the conurbations if effective improvement is to be made. Furthermore, coordinators must have real links with the Public Transport Authorities. In the absence of local highway authority coordination, it is vital that the Regional Offices of the

Department of Transport play this role if the conurbations are to receive their appropriate share of the "national cake" and improve their networks in the public interest.

Road Classification Review

34. Roads are classified in varying ways for different purposes and it is inevitable that some of these approaches will become enshrined in statute and regulation. Without a determined effort to review the situation when circumstances change classifications can be retained beyond their useful life and be a source of some confusion. The improvement of the PRN should be accompanied by a review of both the statutory classifications and the non-mandatory advice carried in the various manuals and notes on highway planning issued by Government with a view to eliminating the redundant and improving coordination. Such a review will bring benefits for the country as a whole and the urban areas in particular.

Conclusions

35. The urban conurbations need improved access to the national motorway network and a good standard PRN is essential for this reason as part of a total urban transport system. This is no soft option because the great urban areas north of London have been so badly hit by the recession with its unemployment and dereliction that they start behind the starting line in the battle for new jobs and urgently need new infrastructure if they are to be competitive. To this end it is important to ensure :-

(a) that the right levels of funding are directed to the conurbations so that the major problems can be tackled in a short timescale and
(b) that the weakened administrative arrangements which will be in place after March 1986 are not allowed to delay progress.

Acknowledgements

36. Special thanks are due to George Parsons, MSc MICE., and other colleagues in the writer's office who have given invaluable help in the preparation of this paper.

References

1. Department of Transport, Transport Statistics, Great Britain, 1974/1984.

2. Transport Bill, 1985.

3. Traffic Signs - Report of the Committee on Traffic Signs for All-Purpose Roads. Worboys Report, 1963.

4. Roads in Urban Areas. HMSO, London, 1966.

CONURBATION ISSUES

5. Standing Conference on London and South East Regional Planning (SRC 3109).

6. Transport in the '80s. County Surveyors' Society, 1981.

7. Press Notice 459. Department of Transport, 16 October, 1984.

8. Department of Transport (Assessment Policy and Methods Division), COBA 9.

9. House of Commons Transport Committee.

Discussion on Papers 10–11

DR D. COOMBE, Halcrow Fox and Associates
With reference to Paper 10, consistency between traffic forecasting and economic evaluation is very important. Current practice often involves using a model to forecast traffic flows and speeds, followed by COBA to calculate the economic benefits. COBA recalculates the speeds and delays, based on the flows output by the traffic model, and often incorrectly as COBA cannot deal with junction interactions, that is, bottlenecks which meter flow downstream and cause queues to form through upstream junctions. It would be better to use one of the new congested assignment models - for example, SATURN or JAM - which calculate delays reasonably well, followed by matrix based economic evaluation using COBA standard values of time, operating costs and so on. The trip matrix can be fixed, or variable to take account of trip generation/suppression (important), peak spreading (important), distribution and modal split (often of little importance as effects of road schemes). COBA, as it stands, has little value as an urban economic evaluation tool, although I find it is of value in the inter-urban scene.

Frameworks are a good way of bringing the effects of urban road schemes together. However, there is a need to avoid bias and double-counting, and to simplify the environmental elements of the framework. For example, inclusion of the economic costs and benefits and net present value can lead to double-counting if the decision-maker is not aware that the former two make up the latter. Some frameworks, especially for schemes in urban areas, have many pages of the environmental effects, often very difficult to assimilate. An extract of main effects, to help discriminate between options, should be prepared in addition to the main framework.

Consideration should be given to making more use of sub-standard grade-separation. In this country, we have a few examples of temporary overpasses, but almost no examples of low headroom underpasses. These low headroom underpasses take up much less room than standard facilities, and can include bends with tight radii. These facilities should be considered carefully as possible solutions at key locations in our urban areas.

CONURBATION ISSUES

MR J. R. ELLIOTT, Greater London Council

In paragraph 3 of Paper 10, Professor Williams states that, 'To ignore the fact that (capacity increases will result in more private traffic) would be irresponsible and myopic)'. I wish to point out the scale of this extra traffic and the consequences on the balance of the overall transport system in London.

The Greater London Council has studied the effects of all the major road enlargements carried out in the London area in the last fifteen years; this is an exercise which the original 1977 Leitch Report said should have been initiated. Two of the results of this review are shown in Figure 1 and Table 1. In Figure 1, separate corridors were analysed which were intended to take account of almost all the reassignment resulting from the road. Table 1 shows the very small amount of reassignment which did take place on the opening of the second Blackwall tunnel. The message emerging from this study is that, where demand has been suppressed by either effective public transport services or a major barrier, the growth of traffic immediately subsequent to the opening of a new road, and up to 2-3 years afterwards, is immense. Traffic growth them slows down as a new point of congestion or transport 'balance' arises.

The rapid growth in traffic in the first years after opening would seem to indicate that it is highly unlikely that the extra traffic is a result of 'redistribution' (change of home or workplace) or from 'development'. This leaves a mode change from public transport as the likely source. This conclusion is backed up by an analysis of the patronage of rail services from the populations of the expanding towns served by the M11. In this case, patronage has declined, whereas for similar towns not served by the M11, rail patronage has increased.

The consequences arising from the effects of new large roads are very serious for the overall transport strategy, as can be illustrated by the resultant vicious circle. More car traffic is indicative of less patronage of public transport; this, under the present FINANCIAL as opposed to ECONOMIC rules for public transport, inevitably leads to higher fares and lower frequency, which ultimately means still more car traffic. A situation could easily arise in which the enlarging of roads would result in a worsening of road travel conditions instead of the planned improvement. It is of utmost importance that, if we are to enlarge the primary route network, much more money is put into the public transport system to maintain or to improve its present competitive position in relation to the private car. I should add that, in a period of two years, the 'Just the Ticket' changes to the fares and ticketing structure in London resulted in 15 per cent fewer car trips by Londoners and a 30 per cent increase in public transport utilisation. Surely a much greater contribution to the improvement of mobility and accessibility then a very elaborate road system.

A 'minor' consequence of the extra traffic is that once

DISCUSSION

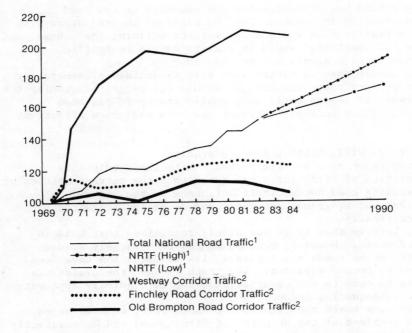

Sources
1. National Road Traffic Forecasts, 1980 and 1984. Department of Transport.
2. GLC Traffic Monitoring Programme.

Figure 1

Table 1. AM peak period traffic flows in the Blackwall Tunnel corridor: 1968-69

	AM peak (0700-0900) Northbound, hourly flows		
	1968	1969	Change
Tower Bridge	1510	1400	-7
Rotherhithe Tunnel	1033	1055	+2
Blackwall Tunnel	1287	2648	+106
Dartford Tunnel	1114	1012	-9
TOTAL	4944	6115	+24

again COBA has overestimated the benefits of new road construction in London. The 'Do minimum' is prejudiced with more traffic than would occur without building the scheme, and the 'Do something' would in reality have more traffic and hence worse congestion than assumed.

I submit that no matter what else is decided elsewhere for the primary route network, we should not suggest expanding the network in London until both public transport has been improved and parking restraint has been made more effective.

MR P. J. WITT, British Road Federation
Every paper by a British author - with the predictable exception of Mr Butler - has questioned the current methods of assessing road needs, their influence on the highway investment programme generally and on design standards specifically.

I believe that it is now widely recognised that COBA is inadequate. However, because it has been largely used on inter-urban roads and bypasses, its inadequacies have been slow to become apparent. The truth is that the closer its results come to our conurbations, the more quickly apparent the inadequacies become.

We have heard the figures for Manchester; we have heard of the problems of the M6 north of Birmingham; and we constantly read about or experience congestion on the M25. The fact is that a system that takes little or no account of traffic generation and economic development is clearly going to underestimate need more often than not. There are also environmental and other subjective judgements.

May I just interject a brief comment here on traffic generation. It is not something to be ashamed of - it is not in some malign way anti-public transport. It is a sign of life and should be welcomed as such. We should all look a little absurd if new roads went largely unused. I know there have been one or two - but we cannot blame COBA for those.

To draw attention to COBA's shortcomings is not of itself a criticism of our colleagues at the DTp. What is a little worrying is the apparent refusal to recognise these shortcomings. When the Secretary of State says publicly that there is nothing wrong with COBA, I think it is a matter for concern.

Professor Williams's Paper underlines the problems only too clearly - perhaps I am being unfair in feeling just a little disappointed that it contained no hint of solutions. We shall have to possess ourselves with patience and await the outcome of his Committee's deliberations. I hope in the end the dificulties will not be allowed to hinder the search for a solution. Man from TM seems to have one option.

In the meantime, our urban problems worsen and too little is being done to put them right. A day so close to 31st March, 1986, is hardly the time to pursue this. We all know that there is no single solution to urban traffic congestion. We

DISCUSSION

need attractive public transport services, better traffic management and more effective law enforcement; and we need better roads to get rid of through traffic and deal with that which has to be there. Environment/Green field/inner cities/regeneration.

If we are to have an effective primary road network, not just lines on a map but roads of appropriate capacity, we must get our assessment methods right. This really needs to be our highest priority in order to prevent the construction of new roads which will not give either road users what they need or everyone a better environment.

DR S. T. ATKINS, Greater London Council

Andrew Sullivan, a leader writer for the Daily Telegraph, has said: "The battle over the GLC revealed that there were policies pursued by Mr Ken Livingstone that were actually popular. His belief that a pleasant urban environment was more important than a purist balance-sheet approach to public transport is to many no more than simple commonsense. Tory local authorities might well learn from those popular decisions in favour of investing in a pleasurable and subsidised transport system, by analysing more seriously proposals for traffic regulation, discouragement of heavy traffic in city or town centres, voucher schemes for private cars, or even an increase in pedestrian zones and cycle lanes. Again, much experience has shown that traditional economic arguments against such improvements to the environment contradict themselves on their own terms: retailing and business benefit from a pleasant shopping environment and from a credible public transport network. Central government might even reassess its rigorous policy of encouraging road against rail, by realising that our towns and cities are being placed under the sort of strain for which they were never designed; that an expansionist motorway system, as well as wrecking areas of natural beauty, may even in the long run encourage the traffic it is designed to alleviate; and that a viable and modernised rail network could prove to be the most cost-effective means of transporting goods and people. A stronger Tory commitment to public transport may, in the present context, be heresy, but it is in line with many popular sentiments, in touch with the particular demands of the English environment, and even economically worthwhile."
(Centre for Policy Studies Report No 72, 1985, p. 41)

The environmentalists have received some unfair and rather glib comment at this conference. I believe it is significant that many of the high-technology industries, whose locational requirements are often independent of traditional constraints, choose to locate in areas of pleasant environmental conditions. A similar effect can be seen within cities, as implied in the previous quotation; a good example is that of the areas adjacent to Covent Garden in London. I believe that it will become accepted that a good physical environment

CONURBATION ISSUES

is a good economic environment.

PROFESSOR WILLIAMS
Dr Coombe's comments on COBA and Framework are certainly relevant, because of the differences which exist between truck road traffic problems in urban and inter-urban locations.

The possibility of double-counting in economic evaluation is recognised as a source of potential errors.

The suggestion that sub-standard grade-separated structures should be considered, for use in urban areas, highlights the importance of traffic management and signs, which are essential operational requirements for roads and traffic in towns.

Mr Elliott refers to the interesting Blackwall Tunnel corridor studies. With his conclusion that a change of mode of transport was the 'likely source' of the extra traffic in question, it is unfortunate that the traffic figures quoted relate only to traffic counts. The results of questionnaire or direct driver interview surveys could have been invaluable in this difficult area of traffic analyses.

Mr Witt's plea for better main roads, traffic management, public transport and more effective enforcement of traffic regulations reflects the obvious need for consistent strategic planning, design and investment policies for road networks and transport systems.

MR MUSTOW
The contributions by Mr Elliott and Dr Atkins of the GLC and Mr Witt of the British Road Federation are important because they highlight both sides of the primary route equation. Urban movement can be understood only in terms of system and it is intellectually indefensible to consider road building and public transport in isolation from each other. The issue is, therefore, one of a balanced provision of roads and public transport in the urban conurbations and not of either one or the other. In this connection, current government attitudes towards urban transport planning are likely to be destructive to any real appreciation or overall transport need.

As a number of speakers have intimated, COBA 9 is not an effective tool for assessing urban road schemes. It is to be hoped that, after its consideration of varying approaches by urban authorities, Professor Williams's Committee will be able to indicate a better way forward.